D1605578

Indiscrete Thoughts

To Arthur Szathmary
with affection and gratitude

Gian-Carlo Rota

Indiscrete Thoughts

Fabrizio Palombi
EDITOR

BIRKHÄUSER
BOSTON • BASEL • BERLIN

Gian-Carlo Rota
Department of Mathematics
Massachusetts Institute of Technology
Cambridge, MA 02139
USA

Editor
Fabrizio Palombi
Department of Philosophy
University of Torino
Torino, Italy

Library of Congress Cataloging-in-Publication Data

Rota, Gian-Carlo, 1932-
Indiscrete thoughts / by Gian-Carlo Rota : edited by Fabrizio Palombi.
p. cm.
Includes bibliographical references and index.
ISBN 0-8176-3866-0 (alk. paper). -- ISBN 3-7643-3866-0 (alk. paper)
1. Mathematics. 2. Sciences. 3. Philosophy. I. Palombi, Fabrizio, 1965- . II. Title
QA7.R65 1996 95-52782
510--dc20 CIP

Printed on acid-free paper

ISBN 0-8176-3866-0
ISBN 3-7643-3866-0
Cover design by Joseph Sherman, Dutton & Sherman Design, Hamden, CT
Typesetting by Hamilton Printing Company, Rensselaer, NY
Printed in the U.S.A.

9 8 7 6 5 4 3 2 1

Contents

Part II. Philosophy: A Minority View

Part III Readings and Comments

Foreword

Reuben Hersh

If you're about to buy this book, you're in for a treat.

I first met Gian-Carlo in the late 70's in Las Cruces, New Mexico. He was there to lecture on his mathematical specialty, combinatorics—how to count complicated finite sets, and extensions of that problem. I repeated inaccurately what I had heard from my honored mentor, Peter Lax of New York University.

"A lecture by Rota is like a double martini!"

It was true. Even to one innocent of combinatorics, his lectures were a delightful combination, both stimulating and relaxing.

There are mathematicians who give great lectures, but write dull, ponderous books. There are some who write beautiful books and give dull lectures. (No names will be mentioned). Gian-Carlo gives brilliant lectures, and his writing is as good. He loves contradiction. He loves to shock. He loves to simultaneously entertain you and make you uncomfortable.

His personal history is rare. He was born in Italy of an architect father who was a leading member of Mussolini's secret hit list. Educated first in Ecuador, then at Princeton and Yale. Started research in a "hot" speciality, functional analysis. Underwent an epiphany and conversion to discrete math—the hard-nosed, down-to-earth stuff—but done from the abstract, high level view point of functional analysis. Developed a major interest in the thought of Husserl, Heidegger and Sartre—phenomenology. With it, a distaste for "analytic philosophy," the ruling trend in Anglo-America.

When he started to teach phenomenology at M.I.T., students wanted "philosophy credit" for his course. The M.I.T. philosophers (analytic, of course) said, "Never," and offered to quit en masse. Jerry

Wiesner, the President, told them, "Go ahead, that will help my budget." They didn't quit. Gian-Carlo's students get philosophy credit.

His large Cambridge apartment has enough bookshelves for many a library, but books are all over everything. I'm not sure he sleeps. Certainly he keeps late hours, reading, writing, doing email, talking on the phone, and thinking.

He has a talent for friendship. Some of my warmest memories are dinners at the home of Stan and Françoise Ulam in Santa Fe, with Gian-Carlo and Mark Kac.

He did me a great service, for which I'd like to thank him again. My first piece on the philosophy of mathematics was rejected, on the advice of a prestigious Harvard philosopher. Gian-Carlo immediately printed it in *Advances in Mathematics*, an absolutely tip-top high-class journal published by Academic Press under the founding editorship of G.-C. Rota. Later he started a second organ, *Advances in Applied Mathematics*. His journals' editorial boards boast the highest conceivable quality and prestige. I once mentioned a difference of opinion with the editor at a journal where I was a board member. Gian-Carlo said, "If anyone on my board gave me trouble, I would kick him out." This book ends with some truly inimitable book reviews from the *Advances*.

Before that come three main parts: *Persons and Places, Philosophy, and Indiscrete Thoughts*.

In *Persons and Places* you'll meet great names. You'll learn not only about their contributions, but also about their kindness, their egos, their absurdities. Most mathematicians know Artin, Church, Feller, Lefschetz and Ulam only as names on a book or paper. Gian-Carlo presents vivid, startling images of them, as live human beings, like us! What a shock! What a liberation! I don't know another since Plutarch who attains this balance: deep appreciation of their accomplishments, and honesty about the embarrassing qualities of great men of science.

After the suave raconteur of *Persons and Places,* Gian-Carlo the phenomenologist may come as a shock. The reader may feel lost for a moment. But, as Richard Nixon said in another connection, "When the going gets tough, the tough get going." Phenomenology is one of the seminal forms of thought of our age. You'll make faster headway with Gian-Carlo than with other writers on the subject. And you can look forward to the easy delights of *Indiscrete Thoughts* coming next.

But you're within your rights if you decide to take phenomenology in small doses. The chapter on mathematical beauty is recommended. Everyone knows that in mathematics, beauty is the highest desideratum. But the few attempts to explain what's meant by "mathematical beauty" have been feeble and unconvincing. Gian-Carlo has a new answer. "Beauty in mathematics is enlightenment." When we are enlightened, we think, "How beautiful." This insight is both enlightening and beautiful!

Indiscrete Thoughts is strong medicine. His thoughts about mathematics are usually startling and provocative. His messages to and about mathematicians are provocative beyond indiscretion. Gian-Carlo never pulls his punches. I advise discretion in reading these indiscretions. A few every few hours. Gulping them at one sitting is not recommended.

What's the thread tying together Gian-Carlo the memoirist, the aphorist, and the phenomenologist? He strives to keep his eyes wide open and then tells it the way he sees it, without pretense, and often without prejudice. Always with wit and flair.

Foreword

Robert Sokolowski

There are two erroneous extremes one might fall into in regard to the philosophy of mathematics. In the one, which we could call naive objectivism, mathematical objects, such as the triangle, the regular solids, the various numbers, a proof, or abelian groups, are taken as simply existent apart from the work of mathematicians. They exist whether we discover them or not. In the other, which we could call naive subjectivism or psychologism, mathematical items are taken to be mental constructs of mathematicians, with no more objectivity than the feelings someone might have had when he thought about a rectangle or worked out a proof. The truth, as usual, lies in the middle. Mathematical items are indeed objective. There are mathematical objects, facts, and valid proofs that transcend the thinking of any individual. However, if we, as philosophers, wish to discuss the objectivity of such items, we must also examine the thinkers, the mathematicians, for whom they are objective, for whom they are facts. The philosophy of mathematics carries out its work by focusing on the correlation between mathematical things and mathematicians.

It is in this correlation, this "in between," that Gian-Carlo Rota has developed his own highly original and programmatic philosophy of mathematics. He draws especially but not exclusively on the phenomenological tradition. The point of Husserl's phenomenology, which was further developed by Heidegger, is that things do appear to us, and we in our consciousness are directed toward them: we are not locked in an isolated consciousness, nor are things mere ciphers that are essentially hidden from us. Rather, the mind finds its fulfillment in the presentation of things, and things are enhanced by the truth they display to us. The subjective is not "merely" subjective but presents objectivity to itself. Husserl's doctrine of the "intentionality" of consciousness breaks through the Cartesian straightjacket that has held so much of modern thought captive.

Another principle in phenomenology is the fact that there are different regions of being, different "eidetic domains," as Rota calls them, and each has its own way of being given to us. Each region calls for a correlative form of thinking that lets the things in it manifest themselves: there are, for example, material objects, living things, human beings, emotional facts, social conventions, economic relationships, and political things, and there are also mathematical items, the domain that Rota has especially explored.

One of the phenomena that he develops in this book is that of "evidence." Most people think that in mathematics truth is reached by proof, specifically by deriving theorems from axioms. Rota shows that such proof is only secondary and derivative. It is not the primary instance of truth. More basic than proof is evidence, which is the self-presentation of a given mathematical object or fact. We can know that some things are true, and we can even know that they must be true, before we have found axiomatic proofs to manifest that truth. We know more than we can prove. Rota shows that axiomatic derivations are ways in which we present mathematical objects and facts, ways in which we try to convey the evidence of the thing in question, ways in which we bring out the possibilities or virtualities of mathematical things. Proofs are valuable not because they bring us assurance that the theorem is in fact true, but because they show the power of the theorem: how the theorem can present itself and hence what it can be.

Evidencing could not occur, of course, except "between" the mathematical object and the mathematician as its dative of presentation, and yet it would be quite incorrect to see evidence as "just" psychological. Rota correctly observes that the depsychologizing of evidence is one of the great achievements of phenomenology. To appeal to terminology used in another philosophical tradition, evidence is the introduction of a fact into the space of reasons, into the domain of logical involvements. Does not such an introduction belong to the space into which it enters? Is not evidence a rational act, indeed, the rational act of the highest order? What is given in evidence is just as logical and rigorous as what is derived within logical space. Mathematical facts are objective but they are achieved and even "owned" by someone in a sense of ownership that is sometimes recognized by adding a person's name to a theorem or conjecture. Rolle's theorem is so named not because of a psychological event but because of an intellectual display, an evidence, a logical event, that took place for someone at a certain time. Once having occurred to him, it can occur again to others at other places and times, and the theorem can be owned by them as well. Intellectual property is not lost when it is given away. Furthermore, we have to be prepared and disposed to let mathematical evidences occur to us: we must live the life of mathematics if we are to see mathematical things.

I think that one of the most valuable moves in this book is Rota's identification of evidence and the Kantian synthetic *a priori*. He observes that "all understanding is synthetic a priori; there is not and there cannot be any other kind." When we achieve evidence, we see something we had not seen before (hence, synthetic), and yet we see that it is necessary, that it could not have been otherwise (hence, *a priori*). Rota's observation sheds light on Kant's theory of judgment, on Husserl's concept of evidence, and on the nature of human understanding.

Another theme developed by Rota is that of "Fundierung." He shows that throughout our experience we encounter things that exist only as founded upon other things: a checkmate is founded upon moving certain pieces of chess, which in turn are founded upon certain pieces of wood or plastic. An insult is founded upon certain words being spoken, an act of generosity is founded upon something's being handed over. In perception, for example, the evidence that occurs to us goes beyond the physical impact on our sensory organs even though it is founded upon it; what we see is far more than meets the eye. Rota gives striking examples to bring out this relationship of founding, which he takes as a logical relationship, containing all the force of logical necessity. His point is strongly antireductionist. Reductionism is the inclination to see as "real" only the foundation, the substrate of things (the piece of wood in chess, the physical exchange in a social phenomenon, and especially the brain as founding the mind) and to deny the true existence of that which is founded. Rota's arguments against reductionism, along with his colorful examples, are a marvelous philosophical therapy for the debilitating illness of reductionism that so pervades our culture and our educational systems, leading us to deny things we all know to be true, such as the reality of choice, of intelligence, of emotive insight, and spiritual understanding. He shows that ontological reductionism and the prejudice for axiomatic systems are both escapes from reality, attempts to substitute something automatic, manageable, and packaged, something coercive, in place of the human situation, which we all acknowledge by the way we live, even as we deny it in our theories.

Rota calls for a widened mathematics that will incorporate such phenomena as evidence and "Fundierung," as well as anticipation, identification, concealment, surprise, and other forms of presentation that operate in our experience and thinking but have not been given an appropriate logical symbolism and articulation. Such phenomena have either not been recognized at all, or they have been relegated to the merely psychological. What has been formalized in logic and mathematics so far have been grammatical operators. It is an exciting and stimulating suggestion to say that various forms of presentation might also be formalized. Rota makes his proposal for a new mathematics in his treatment of artificial intelligence and computer science. These fields, which try to work with intelligent operations wider than those of standard formal logic, have shown, by their failures as well as by their partial successes, that a much richer and more flexible notion of logic is called for. The logic Rota anticipates will not displace the rational animal, the dative of manifestation, but it will bring the power of formalization and mathematics to areas scarcely recognized until now.

Rota's fascinating and sympathetic sketches of persons and places in twentieth-century mathematics should also be seen as part of his study of the correlation between mathematical truth and mathematicians. He sheds light on mathematics by showing the human setting in which it arises. His exhortations to mathematicians to become involved in the service of other disciplines is another point in his recognition of the human face of mathematics. He calls for a presentation of mathematics that uses intuitive, illuminating examples, and for texts with "a discursive, example-rich flow," as opposed to the rigid style that turns the reader into a "code-cracker." The imaginative example is essential to the achievement of mathematical evidence.

Rota makes use of other authors, but never as a mere commentator. He uses authors the way they would most want to be used, as vehicles for getting to the issues themselves. He is like a musician who listens to Mozart and then writes his own music himself. His instinct for mathematical evidence has made him especially alert to philosophical truth. Mathematics and philosophy were blended, after all, in some of the very first philosophers, the Pythagoreans. Their thinking, along with that of all the presocratics, was given a human twist by Socrates, who turned from nature to the human things. Gian-Carlo Rota makes an analogous turn, complementing objective mathematics by showing how it is a human achievement, an intelligent action accomplished by men. His writings have much of Socrates' irony and wit, and the occasional barb is also socratic, meant to illuminate and to sting the reader into looking at things afresh. In these essays, mathematics is restored to its context in being and in human life.

Introduction

Gian-Carlo Rota

The truth offends. In all languages of the world one finds proverbs that stress this truism in many colorful versions (*"veritas odium parit"* in Latin). More precisely: *certain* truths offend. Which truths offend? When and why do we "take offense?"

All cultures have offered variants of one and the same answer to these questions. We take offense at those truths that threaten any of the myths we profess to believe in. Taking offense is an effective way we have of shutting off some unpleasant truth. It works. It enables us to restore a hold on our dearest myths, to last until the next offending truth comes along.

Myths come in two kinds: working myths and wilting myths. Working myths are the bedrock of civilization, they are what college students in the sixties used to call "ultimate reality." We could not function without the solid support that we get from our working myths. We are not aware of our working myths.

Sooner or later, every working myth begins to wilt. We can tell that a myth is wilting as soon as we are able to express it in words. It then turns into a belief, to be preserved and defended.

A wilting myth is an albatross hanging from our necks. Only on rare occasions do we summon the courage to discard a wilting myth; more often, we hang on to a wilting myth to the very end. If anyone dares question any of our wilting myths, we will lash out and label him "elitist," "subversive," "reactionary," "irrational," "cynical," "nihilistic," "obscurantist." We will seize on some incorrect but irrelevant detail as an excuse to dismiss an entire argument. Most discussions, whether in science, in philosophy, in politics or in everyday conversation, are thinly veiled attacks or defenses of some wilting myth.

Eventually, a wilting myth gets dropped by all but the hard-liners.

These are the bigots, the fanatics, the mass murderers. Hitler staged a last-ditch defense of the cloying romantic myths of the past century. Stalin battled for the dying myth of socialism. The kooks of Montana are taking the last stand in defense of the myth of the West.

The wilting myths of the millenium are the theme of this book. Never before in history have so many myths begun to wilt at the same time, and a hard choice had to be made, to wit:

1. *The myth of monolithic personality* "Every scientist must also be a good guy." "If you are good at math, then you will be good at anything." "Great men are great in everything they do." "Heidegger cannot be a good philosopher because he was a Nazi."

Against this myth, sketches of the lives of some notable mathematicians of this century are given in "Fine Hall in the Golden Age," "Light Shadows," "The Story of a Ménage à Trois" and "The Lost Café." When first published, each of these chapters caused a stir of sorts. After reading the section "Problem Solvers and Theorizers," a mathematician friend (one of the most distinguished living mathematicians) wrote that he would not speak to the author ever again. Another mathematician threatened a lawsuit after reading the section on Emil Artin in "Fine Hall in the Golden Age." After publication of a heavily edited version of "The Lost Café" in the magazine *Los Alamos Science*, the author was permanently excluded from the older echelons of Los Alamos society.

2. *The myth of reductionism* "The workings of the mind can be reduced to the brain." "The universe is nothing but a psi function." "Biology is a branch of physics." "Everything has a mechanical explanation."

Critiques of these frequently heard assertions are found in "The Barrier of Meaning," "*Fundierung* as a Logical Concept," "The Primacy of Identity," "The Barber of Seville, or the Useless Precaution," and "Three Senses of 'A is B' in Heidegger." The confusion between scientific thought and reductionist error is rampant in our day, and critiques of reductionism are mistakenly viewed as attacks on the scientific method.

Reductionism would do away with the autonomy of biology and physics, of physics and mathematics, as well as with the autonomy of science and philosophy. "The Pernicious Influence of Mathematics upon Philosophy" is motivated by the loss of autonomy in philosophy. The paper (reprinted five times in four languages) was taken as a personal insult by several living philosophers.

3. *The zero-one myth*. "If a marble is not white, it must be black." "If you don't believe that everything can be explained in terms of atoms and molecules, you must be an irrationalist." "There is no valid explanation other than causal explanation."

The ideal of rationality of the Age of Enlightenment is too narrow, and we need not abandon all reason when we stray from this seventeenth-century straightjacket. Already the life sciences follow a logic that is a long way from the logic of mechanics and causal explanation.

The simplistic cravings for a "nothing but" are dealt with in "The Phenomenology of Mathematical Truth," "The Phenomenology of Mathematical Beauty" and "The Phenomenology of Mathematical Proof." There is no answer to the question "What is mathematics?" because the word "is" is misused in such a question. A distintuished mathematician, who is also one of the last hard-line Stalinists, criticized these essays for their "anarchy." He is right.

The book concludes with a selection of book reviews the author has published in the last twenty-five years. It was hard to resist the temptation to publish samples of the hate mail that was received after these reviews. The truth offends.

The author thanks the editor, Fabrizio Palombi, who organized the text and supplied an ample bibliography, and most of all Ann Kostant of Birkhäuser Boston, without whose help this book would never have seen the light of day.

The author also thanks all readers who have helped with the correction of the galleys: Janis Stipins, Jeff Thompson, Richelle McComas, Barbara and Nick Metas, Peter Ten Eyck, Andrew Wilson, John MacCuish, Michael Hawrylycz, Jeffrey Crants, Krik Krikorian, Daniela Cappelletti, Ottavio D'Antona, Giulio Giorello, Jole Orsenigo, Federico Ponzoni, Giuliano Ladolfi, and many others.

Finally, the author thanks the Sloan Foundation, whose generous grant led to the writing of these essays.

Cambridge, MA, September 1, 1996.

Indiscrete Thoughts

PART I

Persons and Places

CHAPTER I

Fine Hall in its Golden Age

Princeton in the Early Fifties

OUR FAITH IN MATHEMATICS is not likely to wane if we openly acknowledge that the personalities of even the greatest mathematicians may be as flawed as those of anyone else. The greater a mathematician, the more important it is to bring out the contradictions in his or her personality. Psychologists of the future, if they should ever read such accounts, may better succeed in explaining what we, blinded by prejudice, would rather not face up to.

The biographer who frankly admits his bias is, in my opinion, more honest than the one who, appealing to objectivity, conceals his bias in the selection of facts to be told. Rather than attempting to be objective, I have chosen to transcribe as faithfully as I can the inextricable twine of fact, opinion and idealization that I have found in my memories of what happened forty-five years ago. I hope thereby to have told the truth. Every sentence I have written should be prefixed by "It is my opinion that. . . "

I apologize to those readers who may find themselves rudely deprived of the comforts of myth.

Alonzo Church

It cannot be a complete coincidence that several outstanding logicians of the twentieth century found shelter in asylums at some time in their lives: Cantor, Zermelo, Gödel, Peano, and Post are some. Alonzo Church was one of the saner among them, though in some ways his behavior must be classified as strange, even by mathematicians' standards.

He looked like a cross between a panda and a large owl. He spoke softly in complete paragraphs which seemed to have been read out of a book, evenly and slowly enunciated, as by a talking machine. When interrupted, he would pause for an uncomfortably long period to recover the thread of the argument. He never made casual remarks: they did not belong in the baggage of formal logic. For example, he would not say, "It is raining." Such a statement, taken in isolation, makes no sense. (Whether it is actually raining or not does not matter; what matters is consistence). He would say instead, "I must postpone my departure for Nassau Street, inasmuch as it is raining, a fact which I can verify by looking out the window." (These were not his exact words). Gilbert Ryle has criticized philosophers for testing their theories of language with examples which are never used in ordinary speech. Church's discourse was precisely one such example.

He had unusual working habits. He could be seen in a corridor in Fine Hall at any time of day or night, rather like the Phantom of the Opera. Once, on Christmas day, I decided to go to the Fine Hall library (which was always open) to look up something. I met Church on the stairs. He greeted me without surprise.

He owned a sizable collection of science-fiction novels, most of which looked well thumbed. Each volume was mysteriously marked either with a circle or with a cross. Corrections to wrong page numberings in the table of contents had been penciled into several volumes.

His one year course in mathematical logic was one of Princeton University's great offerings. It attracted as many as four students in 1951 (none of them were philosophy students, it must be added, to philosophy's discredit). Every lecture began with a ten-minute ceremony of erasing the blackboard until it was absolutely spotless. We tried to save him the effort by erasing the board before his arrival, but to no avail. The ritual could not be disposed of; often it required water, soap, and brush, and was followed by another ten minutes of total silence while the blackboard was drying. Perhaps he was preparing the lecture while erasing; I don't think so. His lectures hardly needed any preparation. They were a literal repetition of the typewritten text he had written over a period of twenty years, a copy of which was to be found upstairs in the Fine Hall library. (The manuscript's pages had yellowed with the years, and smelled foul. Church's definitive treatise was not published for another five years).[1] Occasionally, one of the sentences spoken in class would be at variance with the text upstairs, and he would warn us in advance of the discrepancy between oral and written presentation. For greater precision, everything he said (except some fascinating side excursions which he invariably prefixed by a sentence like, "I will now interrupt and make a meta-mathematical [sic] remark") was carefully written down on the blackboard, in large English-style handwriting, like that of a grade-school teacher, complete with punctuation and paragraphs. Occasionally, he carelessly skipped a letter in a word. At first we pointed out these oversights, but we quickly learned that they would create a slight panic, so we kept our mouths shut. Once he had to use a variant of a previously proved theorem, which differed only by a change of notation. After a moment of silence, he turned to the class and said, "I could simply say 'likewise,' but I'd better prove it again."

It may be asked why anyone would bother to sit in a lecture which was the literal repetition of an available text. Such a question would betray an oversimplified view of what goes on in a classroom. What one really learns in class is what one does not know at the time one is learning. The person lecturing to us was logic incarnate. His pauses, hesitations, emphases, his betrayals of emotion (however rare) and sundry other nonverbal phenomena taught us a lot more logic than any written text could. We learned to think in unison with him as he spoke, as if following the demonstration of a calisthenics instructor. Church's course permanently improved the rigor of our reasoning.

The course began with the axioms for the propositional calculus (those of Russell and Whitehead's *Principia Mathematica*,[2] I believe) that take material implication as the only primitive connective. The exercises at the end of the first chapter were mere translations of some identities of naive set theory in terms of material implication. It took me a tremendous effort to prove them, since I was unaware of the fact that one could start with an equivalent set of axioms using "and" and "or" (where the disjunctive normal form provides automatic proofs) and then translate each proof step by step in terms of implication. I went to see Church to discuss my difficulties, and far from giving away the easy solution, he spent hours with me devising direct proofs using implication only. Toward the end of the course I brought to him the sheaf of papers containing the solutions to the problems (all problems he assigned were optional, since they could not logically be made to fit into the formal text). He looked at them as if expecting them, and then pulled out of his drawer a note he had just published in *Portugaliae Mathematica*,[3] where similar problems were posed for "conditional disjunction," a ternary connective he had introduced. Now that I was properly trained, he wanted me to repeat the work with conditional disjunction as the primitive connective. His graduate students had declined a similar request, no doubt because they considered it to be beneath them.

Mathematical logic has not been held in high regard at Princeton, then or now. Two minutes before the end of Church's lecture (the course met in the largest classroom in Fine Hall), Lefschetz would begin to peek through the door. He glared at me and the spotless text on the blackboard; sometimes he shook his head to make it clear that he considered me a lost cause. The following class was taught by Kodaira, at that time a recent arrival from Japan, whose work in geometry was revered by everyone in the Princeton main line. The classroom was packed during Kodaira's lecture. Even though his English was atrocious, his lectures were crystal clear. (Among other things, he stuttered. Because of deep-seated prejudices of some of its members, the mathematics department refused to appoint him full-time to the Princeton faculty).

I was too young and too shy to have an opinion of my own about Church and mathematical logic. I was in love with the subject, and his course was my first graduate course. I sensed disapproval all around me; only Roger Lyndon (the inventor of spectral sequences), who had been my freshman advisor, encouraged me. Shortly afterward he himself was encouraged to move to Michigan. Fortunately, I had met one of Church's most flamboyant former students, John Kemeny, who, having just finished his term as a mathematics instructor, was being eased — by Lefschetz's gentle hand — into the philosophy department. (The following year he left for Dartmouth, where he eventually became president).

Kemeny's seminar in the philosophy of science (which that year attracted as many as six students, a record) was refreshing training in basic reasoning. Kemeny was not afraid to appear pedestrian, trivial, or stupid; what mattered was to respect the facts, to draw distinctions even when they clashed with our prejudices, and to avoid black-and-white oversimplifications. Mathematicians have always found Kemeny's common sense revolting.

"There is no reason why a great mathematician should not also be a great bigot," he once said on concluding a discussion whose beginning I have by now forgotten. "Look at your teachers in Fine Hall, at how they treat one of the greatest living mathematicians, Alonzo Church."

I left literally speechless. What? These demi-gods of Fine Hall were not perfect beings? I had learned from Kemeny a basic lesson: a good mathematician is not necessarily a "nice guy."

William Feller

His name was neither William nor Feller. He was named Willibold by his Catholic mother in Croatia, after his birthday saint; his original last name was a Slavic tongue twister, which he changed while still a student at Göttingen (probably on a suggestion of his teacher, Courant). He did not like to be reminded of his Balkan origins, and I had the impression that in America he wanted to be taken for a German who had Anglicized his name. From the time he moved from Cornell to Princeton in 1950, his whole life revolved around a feeling of inferiority. He secretly considered himself to be one of the lowest ranking members of the Princeton mathematics department, probably the second lowest after the colleague who had brought him there, with whom he had promptly quarreled after arriving in Princeton.

In retrospect, nothing could be farther from the truth. Feller's treatise in probability is one of the great masterpieces of mathematics of all time.[4] It has survived unscathed the onslaughts of successive waves of rewriting, and it is still secretly read by every probabilist, many of whom refuse to admit that they still constantly consult it and refer to it as "trivial" (like high school students complaining that Shakespeare's plays are full of platitudes). For a long time, Feller's treatise was the mathematics book most quoted by nonmathematicians.

But Feller would never have admitted to his success. He was one of the first generation who thought probabilistically (the others: Doob, Kac, Lévy, and Kolmogorov), but when it came to writing down any of his results for publication, he would chicken out and recast the mathematics in purely analytic terms. It took one more generation of mathematicians, the generation of Harris, McKean, Ray, Kesten, Spitzer, before probability came to be written the way it is practiced.

His lectures were loud and entertaining. He wrote very large on the blackboard, in a beautiful Italianate handwriting with lots of whirls. Sometimes only one huge formula appeared on the blackboard during the entire period; the rest was handwaving. His proofs — insofar as one can speak of proofs — were often deficient. Nonetheless, they were convincing, and the results became unforgettably clear after he had explained them. The main idea was never wrong.

He took umbrage when someone interrupted his lecturing by pointing out some glaring mistake. He became red in the face and raised his voice, often to full shouting range. It was reported that on occasion he had asked the objector to leave the classroom. The expression "proof by intimidation" was coined after Feller's lectures (by Mark Kac). During a Feller lecture, the hearer was made to feel privy to some wondrous secret, one that often vanished by magic as he walked out of the classroom at the end of the period. Like many great teachers, Feller was a bit of a con man.

I learned more from his rambling lectures than from those of anyone else at Princeton. I remember the first lecture of his I ever attended. It was also the first mathematics course I took at Princeton (a course in sophomore differential equations). The first impression he gave me was one of exuberance, of great zest for living, as he rapidly wrote one formula after another on the blackboard while his white mane floated in the air. After the first lecture, I had learned two words which I had not previously heard: "lousy" and "nasty." I was also terribly impressed by a trick he explained: the integral $\int_0^{2\pi} cos^2\, x\, dx$ equals the integral $\int_0^{2\pi} sin^2\, x\, dx$ and therefore, since the sum of the two integrals equals 2π, each of them is easily computed.

He often interrupted his lectures with a tirade from the repertoire he had accumulated over the years. He believed these side shows to be a necessary complement to the standard undergraduate curriculum. Typical titles: "Ghandi was a phony," "Velikovsky is not as wrong as you think," "Statisticians do not know their business," "ESP is a sinister plot against civilization," "The smoking and health report is all wrong." Such tirades, it must be said to his credit, were never repeated to the same class, though they were embellished with each performance. His theses, preposterous as they sounded, invariably carried more than an element of truth.

He was Velikovsky's next-door neighbor on Random Road. They first met one day when Feller was working in his garden pruning some bushes, and Velikovsky rushed out of his house screaming, "Stop! You are killing your father!" Soon afterward they were close friends.

He became a crusader for any cause which he thought to be right, no matter how orthogonal to the facts. Of his tirades against statistics,

I remember one suggestion he made in 1952, which still appears to me to be quite sensible: in multiple-choice exams, students should be asked to mark one wrong answer, rather than to guess the right one. He inveighed against American actuaries, pointing to Swedish actuaries (who gave him his first job after he graduated from Göttingen) as the paradigm. He was so vehemently opposed to ESP that his overkill (based on his own faulty statistical analyses of accurate data) actually helped the other side.

He was, however, very sensitive to criticism, both of himself and of others. "You should always judge a mathematician by his best paper!" he once said, referring to Richard Bellman.

While he was writing the first volume of his book he would cross out entire chapters in response to the slightest critical remark. Later, while reading galleys, he would not hesitate to rewrite long passages several times, each time using different proofs; some students of his claim that the entire volume was rewritten in galleys, and that some beautiful chapters were left out for fear of criticism. The treatment of recurrent events was the one he rewrote most, and it is still, strictly speaking, wrong. Nevertheless, it is perhaps his greatest piece of work. We are by now so used to Feller's ideas that we tend to forget how much mathematics today goes back to his "recurrent events;" the theory of formal grammars is one outlandish example.

He had no firm judgment of his own, and his opinions of other mathematicians, even of his own students, oscillated wildly and frequently between extremes. You never knew how you stood with him. For example, his attitude toward me began very favorably when he realized I had already learned to differentiate and integrate before coming to Princeton. (In 1950, this was a rare occurrence). He all but threw me out of his office when I failed to work on a problem on random walk he proposed to me as a sophomore; one year later, however, I did moderately well on the Putnam Exam, and he became friendly again, only to write me off completely when I went off to Yale to study functional analysis. The tables were turned again in 1963 when he gave me a big hug at a meeting of the AMS in New York. (I learned shortly afterward that Doob had explained to him my 1963 limit theorem for positive operators. He liked the ideas of "strict sense spectral theory" so much that he invented the phrase "to get away with Hilbert space"). His benevolence, alas, proved to be short-lived: as soon as I started working in combinatorics, he stopped talking to me. But not, fortunately, for long: he listened to a lecture of mine on applications of exterior algebra to combinatorics and started again singing my praises to everyone. He had jumped to the conclusion that I was the inventor of exterior algebra. I never had the heart to tell him the truth. He died believing I was the latter-day Grassmann.

He never believed that what he was doing was going to last long, and he modestly enjoyed pointing out papers that made his own work obsolete. No doubt he was also secretly glad that his ideas were being kept alive. This happened with the Martin boundary ("It is so much better than *my* boundary!") and with the relationship between diffusion and semigroups of positive operators.

Like many of Courant's students, he had only the vaguest ideas of any mathematics that was not analysis, but he had a boundless admiration for Emil Artin and his algebra, for Otto Neugebauer and for German mathematics. Together with Emil Artin, he helped Neugebauer figure out the mathematics in cuneiform tablets. Their success gave him a new harangue to add to his repertoire: "The Babylonians knew Fourier analysis." He would sing the praises of Göttingen and of the Collège de France in rapturous terms. (His fulsome encomia of Europe reminded me of the sickening old Göttingen custom of selling picture postcards of professors). He would tell us bombastic stories of his days at Göttingen, of his having run away from home to study mathematics (I never believed that one), and of how, shortly after his arrival in Göttingen, Courant himself visited him in his quarters while the landlady watched in awe.

His views on European universities changed radically after he made a lecture tour in 1954; from that time on, he became a champion of American know-how.

He related well to his superiors and to those whom he considered to be his inferiors (such as John Riordan, whom he used to patronize), but his relations with his equals were uneasy at best. He was particularly harsh with Mark Kac. Kitty Kac once related to me an astonishing episode. One summer evening at Cornell Mark and Kitty were sitting on the Fellers' back porch in the evening. At some point in the conversation, Feller began a critique of Kac's work, paper by paper, of Kac's working habits, and of his research program. He painted a grim picture of Kac's future, unless Mark followed Willy's advice to master more measure theory and to use almost-everywhere convergence rather than the trite (to Willy) convergence in distribution. As Kitty spoke to me — a few years after Mark's death, with tears in her eyes — I could picture Feller carried away by the sadistic streak that emerges in our worst moments, when we tear someone to shreds with the intention of forgiving him the moment he begs for mercy.

I reassured Kitty that the Feynman-Kac formula (as Jack Schwartz named it in 1955) will be remembered in science long after Feller's book is obsolete. I could almost hear a sigh of relief, forty-five years after the event.

Emil Artin

Emil Artin came to Princeton from Indiana shortly after Wedderburn's death in 1946. Rumor had it (*se non è vero è ben trovato*) that Indiana University had decided not to match the Princeton offer, since during the ten years of his research tenure he had published only one research paper, a short proof of the Krein-Milman theorem in "The Piccayune [sic] Sentinel," Max Zorn's *samizdat* magazine.

A few years later, Emil Artin had become the idol of Princeton mathematicians. His mannerisms did not discourage the cult of personality. His graduate students would imitate the way he spoke and walked, and they would even dress like him. They would wear the same kind of old black leather jacket he wore, like that of a Luftwaffe pilot in a war movie. As he walked, dressed in his too-long winter coat, with a belt tightened around his waist, with his light blue eyes and his gaunt face, the image of a Wehrmacht officer came unmistakably to mind. (Such a military image is wrong, I learned years later from Jürgen Moser. Germans see the Emil Artin "type" as the epitome of a period of Viennese *Kultur*).

He was also seen wearing sandals (like those worn by Franciscan friars), even in cold weather. His student Serge Lang tried to match eccentricities by never wearing a coat, although he would always wear heavy gloves every time he walked out of Fine Hall, to protect himself against the rigors of winter.

He would spend endless hours in conversation with his few protégés (at that time, Lang and Tate), in Fine Hall, at his home, during long walks, even via expensive long-distance telephone calls. He spared no effort to be a good tutor, and he succeeded beyond all expectations.

He was, on occasion, tough and rude to his students. There were embarrassing public scenes when he would all of a sudden, at the most unexpected times, lose his temper and burst into a loud and unseemly "I told you a hundred times that. . . " tirade directed at one of them. One of these outbursts occurred once when Lang loudly proclaimed that Pólya and Szegö's problems were bad for mathematical education. Emil Artin loved special functions and explicit computations, and he relished Pólya and Szegö's *Aufgaben und Lehrsätze*,[5] though his lectures were the negation of any anecdotal style.

He would also snap back at students in the honors freshman calculus class which he frequently taught. He might throw a piece of chalk or a coin at a student who had asked too silly a question ("What about the null set?"). A few weeks after the beginning of the fall term, only the bravest would dare ask any more questions, and the class listened in sepulchral silence to Emil Artin's spellbinding voice, like a congregation at a religious service.

He had definite (and definitive) views on the relative standing of most fields of mathematics. He correctly foresaw and encouraged the rebirth of interest in finite groups that was to begin a few years later with the work of Feit and Thompson, but he professed to dislike semigroups. Schützenberger's work, several years after Emil Artin's death, has proved him wrong: the free semigroup is a far more interesting object than the free group, for example. He inherited his mathematical ideas from the other great German number theorists since Gauss and Dirichlet. To all of them, a piece of mathematics was the more highly thought of, the closer it came to Germanic number theory.

This prejudice gave him a particularly slanted view of algebra. He intensely disliked Anglo-American algebra, the kind one associates with the names of Boole, C. S. Peirce, Dickson, the late British invariant theorists (like D. E. Littlewood, whose proofs he would make fun of), and Garrett Birkhoff's universal algebra (the word "lattice" was expressly forbidden, as were several other words). He thought this kind of algebra was "no good" — rightly so, if your chief interests are confined to algebraic numbers and the Riemann hypothesis. He made an exception, however, for Wedderburn's theory of rings, to which he gave an exposition of as yet unparalleled beauty.

A great many mathematicians in Princeton, too awed or too weak to form opinions of their own, came to rely on Emil Artin's pronouncements like hermeneuts on the mutterings of the Sybil at Delphi. He would sit at teatime in one of the old leather chairs ("his" chair) in Fine Hall's common room and deliver his opinions with the abrupt definitiveness of Wittgenstein's or Karl Kraus's aphorisms. A gaping crowd of admirers and worshippers, often literally sitting at his feet, would record them for posterity. Sample quips: "If we knew *what* to prove in non-Abelian class field theory, we could prove it;" "Witt was a Nazi, the one example of a clever Nazi" (one of many exaggerations). Even the teaching of undergraduate linear algebra carried the imprint of Emil Artin's very visible hand: we were to stay away from any mention of bases and determinants (a strange injunction, considering how much he liked to compute). The alliance of Emil Artin, Claude Chevalley, and André Weil was out to expunge all traces of determinants and resultants from algebra. Two of them are now probably turning in their graves.

His lectures are best described as polished diamonds. They were delivered with the virtuoso's spontaneity that comes only after length and excruciating rehearsal, always without notes. Very rarely did he make a mistake or forget a step in a proof. When absolutely lost, he would pull out of his pocket a tiny sheet of paper, glance at it quickly, and then turn to the blackboard, like a child caught cheating.

He would give as few examples as he could get away with. In a course in point-set topology, the only examples he gave right after defining the notion of a topological space were a discrete space and an infinite set with the finite-cofinite topology. Not more than three or four more examples were given in the entire course.

His proofs were perfect but not enlightening. They were the end results of years of meditation, during which all previous proofs of his and of his predecessors were discarded one by one until he found the definitive proof. He did not want to admit (unlike a wine connoisseur, who teaches you to recognize *vin ordinaire* before allowing you the *bonheur* of a *premier grand cru*) that his proofs would best be appreciated if he gave the class some inkling of what they were intended to improve upon. He adamantly refused to give motivation of any kind in the classroom, and stuck to pure concepts, which he intended to communicate *directly*. Only the very best and the very worst responded to such shock treatment: the first because of their appreciation of superior exposition, and the second because of their infatuation with Emil Artin's style. Anyone who wanted to understand had to figure out later "what he had really meant."

His conversation was in stark contrast to the lectures: he would then give out plenty of relevant and enlightening examples, and freely reveal the hidden motivation of the material he had so stiffly presented in class.

It has been claimed that Emil Artin inherited his flair for public speaking from his mother, an opera singer. More likely, he was driven to perfection by a firm belief in axiomatic *Selbständigkeit*. The axiomatic method was only two generations old in Emil Artin's time, and it still had the force of a magic ritual. In his day, the identification of mathematics with the axiomatic method for the presentation of mathematics was not yet thought to be a preposterous misunderstanding (only analytic philosophers pull such goofs today). To Emil Artin, axiomatics was a useful technique for disclosing hidden analogies (for example, the analogy between algebraic curves and algebraic number fields, and the analogy between the Riemannian hypothesis and the analogous hypothesis for infinite function fields, first explored in Emil Artin's thesis and later generalized into the "Weil conjectures"). To lesser minds, the axiomatic method was a way of grasping the "modern" algebra that Emmy Noether had promulgated, and that her student Emil Artin was the first to teach. The table of contents of every algebra textbook is still, with small variations, that which Emil Artin drafted and which van der Waerden was the first to develop. (How long will it take before the imbalance of such a table of contents — for example, the overemphasis on Galois theory at the expense of tensor algebra — will be recognized and corrected)?

At Princeton, Emil Artin and Alonzo Church inspired more loyalty in their students than Bochner or Lefschetz. It is easy to see why. Both of them were prophets of new faiths, of two conflicting philosophies of algebra that are still vying with each other for mastery.

Emil Artin's mannerisms have been carried far and wide by his students and his students' students, and are now an everyday occurrence (whose origin will soon be forgotten) whenever an algebra course is taught. Some of his quirks have been overcompensated: Serge Lang will make a *volte-face* on any subject, given adequate evidence to the contrary; Tate makes a point of being equally fair to all his doctoral students; and Arthur Mattuck's lectures are an exercise in high motivation. Even his famous tantrums still occur. A few older mathematicians still recognize in the outbursts of the students the gestures of the master.

Solomon Lefschetz

No one who talked to Lefschetz failed to be struck by his rudeness. I met him one afternoon at tea, in the fall term of my first year at Princeton, in the Fine Hall common room. He asked me if I was a graduate student; after I answered in the negative, he turned his back and behaved as if I did not exist. In the spring term, he suddenly began to notice my presence. He even remembered my name, to my astonishment. At first, I felt flattered, until (perhaps a year later) I realized that what he remembered was not me, but the fact that I had an Italian name. He had the highest regard for the great Italian algebraic geometers, for Castelnuovo, Enriques, and Severi, who were slightly older than he was, and who were his equals in depth of thought as well as in sloppiness of argument. "You should have gone to school in Rome in the twenties. That was the Princeton of its time!" he told me.

He was rude to everyone, even to people who doled out funds in Washington and to mathematicians who were his equals. I recall Lefschetz meeting Zariski, probably in 1957 (while Hironaka was already working on the proof of the resolution of singularities for algebraic varieties). After exchanging with Zariski warm and loud Jewish greetings (in Russian), he proceeded to proclaim loudly (in English) his skepticism on the possibility of resolving singularities for all algebraic varieties. "Ninety percent proved is zero percent proved!" he retorted to Zariski's protestations, as a conversation stopper. He had reacted similarly to several other previous attempts that he had to shoot down. Two years later he was proved wrong. However, he had the satisfaction of having been wrong only once.

He rightly calculated that skepticism is always a more prudent policy when a major mathematical problem is at stake, though it did not occur to him that he might express his objections in less obnoxious language. When news first came to him from England of Hodge's work on harmonic integrals and their relation to homology, he dismissed it as the work of a crackpot, in a sentence that has become a proverbial mathematical gaffe. After that débacle, he became slightly more cautious.

Solomon Lefschetz was an electrical engineer trained at the École Centrale, one of the lesser of the French *grandes écoles*. He came to America probably because, as a Russian-Jewish refugee, he had trouble finding work in France. A few years after arriving in America, an accident deprived him of the use of both hands. He went back to school and got a quick Ph.D. in mathematics at Clark University (which at that time had a livelier graduate school than it has now). He then accepted instructorships at the Universities of Nebraska and Kansas, the only means he had to survive. For a few harrowing years he worked night and day, publishing several substantial papers a year in topology and algebraic geometry. Most of the ideas of present-day algebraic topology were either invented or developed (following Poincaré's lead) by Lefschetz in these papers; his discovery that the work of the Italian algebraic geometers could be recast in topological terms is only slightly less dramatic.

To no one's surprise (except that of the anti-Semites who still ruled over some of the Ivy League universities), he received an offer to join the Princeton mathematics department from Luther Pfahler Eisenhart, the chairman, an astute mathematician whose contributions to the well-being of mathematics have never been properly appreciated (to his credit, his books, carefully and courteously written as few mathematics books are, are still in print today).

His colleagues must have been surprised when Lefschetz himself started to develop anti-Semitic feelings which were still lingering when I was there. One of the first questions he asked me after I met him was whether I was Jewish. In the late thirties and forties, he refused to admit any Jewish graduate students in mathematics. He claimed that, because of the Depression, it was too difficult to get them jobs after they earned their Ph.D.'s. He liked and favored red-blooded American boyish Wasp types (like Ralph Gomory), especially those who came from the sticks, from the Midwest, or from the South.

He considered Princeton to be a just reward for his hard work in Kansas, as well as a comfortable, though only partial, retirement home. After his move he did little new work of his own in mathematics, though he did write several books, among them the first comprehensive treatise on topology. This book, whose influence on the further development of the subject was decisive, hardly contains one completely correct proof. It was rumored that it had been written during one of Lefschetz's sabbaticals away from Princeton, when his students did not have the opportunity to revise it and eliminate the numerous errors, as they did with all of their teacher's other writings.

He despised mathematicians who spent their time giving rigorous or elegant proofs for arguments which he considered obvious. Once, Spencer and Kodaira, still associate professors, proudly explained to him a clever new proof they had found of one of Lefschetz's deeper theorems. "Don't come to me with your pretty proofs! We don't bother with that baby stuff around here!" was his reaction. Nonetheless, from that moment on he held Spencer and Kodaira in high esteem. He liked to repeat, as an example of mathematical pedantry, the story of one of E. H. Moore's visits to Princeton, when Moore started a lecture by saying, "Let a be a point and let b be a point." "But why don't you just say, 'Let a and b be points!'" asked Lefschetz. "Because a may equal b," answered Moore. Lefschetz got up and left the lecture room.

Lefschetz was a purely intuitive mathematician. It was said of him that he had never given a completely correct proof, but had never made a wrong guess either. The diplomatic expression "open reasoning" was invented to justify his always deficient proofs. His lectures came close to incoherence. In a course on Riemann surfaces, he started with a string of statements in rapid succession, without writing on the blackboard: "Well, a Riemann surface is a certain kind of Hausdorff space. You know what a Hausdorff space is, don't you? It is also compact, o.k.? I guess it is also a manifold. Surely you know what a manifold is. Now let me tell you one nontrivial theorem: the Riemann-Roch Theorem." And so on until all but the most faithful students dropped out of the course.

I listened to a few of his lectures, curious to find out what he might be saying in a course on ordinary differential equations he had decided to teach on the spur of the moment. He would be holding a piece of chalk with his artificial hands, and write enormous letters on the blackboard, like a child learning how to write. I could not make out the sense of anything he was saying, nor whether what he was saying was gibberish to me alone or to everyone else as well. After one lecture, I asked a rather senior-looking mathematician who had been religiously attending Lefschetz's lectures whether he understood what the lecturer was talking about. I received a vague and evasive answer. After that moment, I knew.

When he was forced to relinquish the chairmanship of the Princeton mathematics department for reasons of age, he decided to promote Mexican mathematics. His love/hate of the Mexicans got him into trouble. Once, in a Mexican train station, he spotted a charro dressed in full regalia, complete with a pair of pistols and rows of cartridges across his chest. He started making fun of the charro's attire, adding some deliberate slurs in his excellent Spanish. His companions feared that the charro might react the way Mexicans traditionally react to insult. The charro eventually stood up and reached for his pistols. Lefschetz looked at him straight in the face and did not back off. There were a few seconds of tense silence. "Gringo loco!" said the charro finally, and walked away. When Lefschetz decided to leave Mexico and come back to the United States, the Mexicans awarded him the Order of the Aztec Eagle.

During Lefschetz's tenure as chairman of the mathematics department, Princeton became the world center of mathematics. He had an uncanny instinct for sizing up mathematicians' abilities, and he was invariably right when sizing up someone in a field where he knew next to nothing. In topology, however, his judgment would slip, probably because he became partial to work that he half understood.

His standards of accomplishment in mathematics were so high that they spread by contagion to his successors, who maintain them to this day. When addressing an entering class of twelve graduate students, he told them in no uncertain terms, "Since you have been carefully chosen among the most promising undergraduates in mathematics in the country, I expect that you will all receive your Ph.D.'s rather sooner than later. Maybe one or two of you will go on to become mathematicians."

CHAPTER II

Light Shadows

Yale in the Early Fifties

To Jack Schwartz on his sixty-fifth birthday

Jack Schwartz

If a twentieth century version of Emerson's *Representative Men*[1] were ever to be written, Jack Schwartz would be the subject of one of the chapters. The achievements in the exact sciences of the period that runs from roughly 1930 to 1990 may well remain unmatched in any foreseeable future. Jack Schwartz' name will be remembered as a beacon of this age.

No one among the living has left as broad and deep a mark in as many areas of pure and applied mathematics, in computer science, economics, physics, as well as in fields which ignorance prevents me from naming.

I'd like to recall a few anecdotes from a brief period of the past, the years 1953 to 1955, when I met Jack and learned mathematics as a graduate student at Yale.

The first lecture by Jack I listened to was given in the spring of 1954 in a seminar in functional analysis. A brilliant array of lecturers had been expounding throughout the spring term on their pet topics. Jack's lecture dealt with stochastic processes. Probability was still a mysterious subject cultivated by a few scattered mathematicians, and the expression "Markov chain" conveyed more than a hint of mystery. Jack started his lecture with the words, "A Markov chain is a generalization of a function." His perfect motivation of the Markov property put the audience at ease. Graduate students and instructors relaxed and followed his every word to the end.

Jack's sentences are lessons in clarity and poise. I remember a discussion in the mid-eighties about the future of artificial intelligence in which for some reason I was asked to participate. The advocates of what was then called "hard A.I.," to the dismay of their opponents, were painting a triumphalist picture of the future of computer intelligence. As the discussion went on, all semblance of logical argument was given up. Eventually, everyone realized that Jack had not said a word, and all faces turned toward him. "Well," he said after a pause, "some of these developments may lie one hundred Nobel prizes away." His felicitous remark had a calming effect. The A.I. people felt they were being granted the scientific standing they craved, and their opponents felt vindicated by Jack's assertion.

I have made repeated use in my own lectures of Jack's strikingly appropriate phrases. You may forgive this shameless appropriation upon learning that my students have picked up the very same phrases from me.

From Princeton to Yale

Mathematics in the fifties was a marginal subject, like Latin and Greek. The profession of mathematician had not yet been recognized by the public, and it was not infrequent that a mathematics graduate student might be asked whether he was planning to become an actuary. The centers of mathematics were few and far between, and communication among them was infrequent. The only established departments were Princeton and Chicago. Harvard was a distant third, and Yale was in the process of overcoming its dependence on the College. In New York, Richard Courant was busy setting up his Institute of Mathematical Sciences at 25 Waverly Place, having just finished training his first generation of students in America: Lax and Nirenberg, Cathleen Morawetz and Harold Grad. It was clear that the Institute he was assembling was going to become a great center of mathematics.

In the spring of 1953 I was a senior at Princeton, and I had applied to various universities for admission to graduate school. It soon became apparent that I only needed to apply to one graduate school. Professor A. W. Tucker was not yet the chairman of the mathematics department, but he was already acting as if he were. Solomon Lefschetz, the nominal chairman on the verge of retirement, used to make fun of Tucker by lavishing in public uncomfortably high praise of Tucker's managerial skills.

There were few undergraduate majors, maybe a half dozen each year, and Al Tucker saw to it that they were sent to the "right" graduate schools. He made sure that Jack Milnor stayed at Princeton, and he sent Hyman Bass, Steve Chase, and Jack Eagon to Chicago, Mike Artin to Harvard. In April 1953, I wrote a letter of acceptance to the University of Chicago, which had offered me a handsome fellowship (in those days, it was extremely easy to be offered a graduate fellowship anywhere). On my way to the mailbox, I met Professor Tucker on the narrow, rickety stairs of the old Fine Hall. He asked me where I had decided to go to graduate school, and upon hearing of my decision, he immediately retorted, "You are not going to Chicago, you are going to Yale!"

Since he was one of my professors, I had no choice but to do his bidding. I tore up the letter to Chicago and wrote an identical letter of acceptance of the fellowship I had been offered by Yale.

In retrospect, my decision to go to Yale was one of the few right decisions I have made, and I will always thank Al Tucker's guiding hand. Don Spencer, another of my undergraduate teachers, was first to mention the name of Jack Schwartz to me. He complimented me on my choice of graduate school, remarking, "Oh yes, Yale, that is where Jack Schwartz is. . . ." It was an astounding statement, considering that Jack Schwartz was getting his Ph.D. from Yale that very month. Spencer's comment began an irreversible process of turning Jack Schwartz into a mythological figure in my mind, a process that did not stop after I met him a few months later. I have never been able to stop the process.

Josiah Willard Gibbs

The sciences at Yale have always played second fiddle to the humanities. At faculty meetings it is not unusual to witness a professor of literature point with a wide gesture, like a Roman senator, towards Hillhouse Avenue, where most of the science departments are located, and begin an oratorical sentence with the words, "Even in the sciences. . . ."

Despite the distrust of science, Yale College was once blessed with the presence of one of the foremost scientists of the nineteenth century, Josiah Willard Gibbs. Gibbs served as a professor at Yale without any stipend. Professors seldom received any salary from the university in those days. Teaching young men from the upper echelons was not a salaried profession but a privilege for the happy few. The administration did, of course, receive handsome salaries, like all administrations of all times.

One day Gibbs received an offer from the recently founded Johns Hopkins University. It was an endowed professorship. It was probably the position relinquished by Sylvester who had accepted a professorship at Oxford after religious vows for professors were dropped as a requirement.

Thanks to its endowment, the Johns Hopkins professorship carried an annual stipend of one hundred dollars. It is unclear whether Gibbs was delighted with the offer; in any case he felt obliged to get ready to move to Baltimore. One of his colleagues, realizing that Gibbs was packing up, hastened to contact the Dean of the College. The Dean then asked the colleague if he could do something to keep Gibbs at Yale. "Why, just tell him that you'd like him to remain at Yale!" he answered.

The Dean kept his word and did what had been recommended. He summoned Gibbs to his office and let him know that he unquestionably wanted Gibbs to stay. It was the kind of reassurance Gibbs needed. He declined the Johns Hopkins offer, and remained at Yale for the rest of his career.

Some of Gibbs' most original papers in statistical mechanics were published in the Proceedings of the Connecticut Academy of Sciences. One might wonder how papers which saw the light in such an obscure publication could manage to receive worldwide publicity and acclaim within so short a time. After I moved to Yale in the summer of 1953, I accidentally found the answer.

In the fifties Yale had no mathematics library. Math books were relegated to a few shelves in Sterling Library, randomly classified under that miscarriage of reason that was the Dewey Decimal System. I recall that in August of 1953 I used to walk through the mathematics stacks of the Sterling Library and pull out books from here and there, as we do in childhood. All students had access to the shelves. (Not until the early sixties was a mathematics library created, after several members of the mathematics department had threatened to quit).

Next to an array of perused calculus books was a set of hardbound lecture notes of courses offered at Yale at various times by members of the faculty. Among these were some course notes by Gibbs, presumably hand-written. A few additional sheets were glued to one of these volumes, listing the names of all the notable scientists of Gibbs' time: from Poincaré, Hilbert, Boltzmann and Mach, to individuals who are now all but forgotten. Altogether, more than two hundred names and addresses were alphabetized in a beautiful, faded handwriting. Those sheets were a copy of Gibbs' mailing list. Leafing through in amazement, I realized at last how Gibbs had succeeded in getting himself to be known in a short time. I also learned an instant lesson: the importance of keeping a mailing list.

Yale in the Fifties

In the early fifties Yale had not yet lost the charm of a posh out-of-the-way college for the children of the wealthy. Erwin Chargaff, in his autobiography *Heraclitean Fire*, describes Yale as follows:

> Yale University was much more of a college than a graduate school; and the undergraduates were all over town. They were digesting their last goldfish, for the period of whoopee, speakeasies, and raccoon coats was coming to an end, to be replaced by a grimmer America which was never to recover the jóy of upper-class life. The University proper was much less in evidence. Shallow celebrities, such as William Lyon Phelps, owed their evanescent fame to the skill with which they kept their students in a state of elevated somnolence.[2]

The main part of the campus, consisting of nine shining colleges in the middle of New Haven, was of recent vintage. At the lower end of Hillhouse Avenue, the red bricks of Silliman College shone like the plaster of a movie set as one made one's way back to the main campus from the deliberately distant science buildings. Envious Englishmen spread the malicious rumor that the colleges built with Mr. Harkness' money were Hollywood-style imitations of Oxford colleges. But nowadays the shoe is on the other foot, and it is Oxford that is at the receiving end of other jibes.

The graduate school was a genteel, though no longer gentile, appendage added to the University by gracious assent of the Dean of the College. It was this Dean who held the real power and could overrule the President. Since the thirties, professors appointed to the few and ill-paid graduate chairs had consistently turned out to be better scholars and scientists than the administration had foreseen at tenure time. Nonetheless, evil tongues from northern New England whispered that a certain well-known physics professor would never have made it past assistant professor in Cambridge; but he was one of the last exceptions, soon to fade into bestsellerdom.

Hard work, the kind one reads about in the hagiographies of scientists, was regarded by the graduate students with embarrassment. It was not unusual for a graduate student to spend seven postgraduate years as a teaching assistant before being reluctantly awarded a terminal Ph.D. The university cynically encouraged graduate students to defer their degrees: the money saved by hiring low-paid teaching assistants in place of professors could be used to enrich the rare book collection. Writing a doctoral dissertation was an in-house affair, having little to do with publishing or distasteful professionalism.

On learning about the shocking leisure of graduate life at Yale in the fifties, one may seek shelter in one of the current philosophies of education that promise instant relief from the onslaught of reality. One would thereby be led to the mistaken conclusion that "creativity" (a pompous word currently enjoying a fleeting but insidious vogue) would be stifled in the constricted, provincial, unhurried atmosphere of New Haven. The facts tell a different story. The comforts of an easy daily routine in a rigidly circumscribed environment, encouraged by the indulgent scrutiny of benign superiors, foster the life of the mind. Professors were poorly paid but enjoyed unquestioned prestige. In their sumptuous quarters in the colleges they could encourage their students with sherry and conversation.

Purposeless delectation in ideas may be as educational as intensive study. At Yale, the enjoyment of an absorbing range of campus activities was combined with the lingering belief that nothing much mattered in that little corner of the world. Teachers and students were thereby led to meet the principal requirement of a successful educational experience: they were kept from taking themselves too seriously.

I cannot help but recall one of the episodes of my graduate career at Yale. While I was a second-year graduate student, Mr. Holden, the secretary of the University, was charged with finding a suitable escort for Mrs. William Sloane Coffin, Sr., one of Yale's most generous benefactors. After extensive consultations with the director of International House, Mr. Holden asked me whether I would be willing to do the job, and I accepted.

I will never forget the evenings I escorted Mrs. Coffin to the concerts. On the second Wednesday of each month, I believe, the Music School sponsored a concert in Woolsey Hall. Her big black limousine was the only car allowed on campus. I used to wait for her at the bottom of the steps of Woolsey Hall, dressed in my father's pre-war dark suit. The uniformed driver would help her out, and she would reach for my shoulder; together, we would slowly start up the broad white staircase towards the glittering concert hall. She was almost blind, and she leaned heavily on me as we climbed. As we entered the Woolsey Hall rotunda, a high officer of the University would suddenly appear and rush over to greet her, brushing me aside. I would walk her around again during intermissions, amid the smiles of Connecticut gentry. It was like a scene from "Gone with the Wind."

The few mathematicians who attended the concert were unaware of the nature of my relationship with Mrs. Coffin. Naive as always, the younger ones (Henry Helson, Larry Marcus, John Wermer) used to listen in seriousness to my improvised answers to Mrs. Coffin's inquiries about the bridges on the Arno in Florence (a city where I had never been) or about the poetry of Ludovico Ariosto. On the latter topic I felt I could give ampler explanations since no translation of Ariosto was available in English.

The following year my Sheffield Fellowship was upgraded to a Sterling Fellowship.

There is a fundamental difference between the quality of life in Northern New England and Southern New England. It comes from the shadows. On a sunny day in Cambridge, the sharp shadows across the Charles River deepen the blue hues of the water and cut out the outlines of the distant buildings of Boston as if fashioned from stiff cardboard. In New Haven, by contrast, the light shadows are softened in a silky white haze, which encloses the colleges in a cozy aura of unreality. Such foresight of Mother Nature bespeaks a parting of destinies.

Mathematics at Yale

The Yale mathematics department was the first of the science departments to awaken. It was not until the fifties when the last of a long line of professional teachers of calculus retired: fine, upright gentlemen of the old school, richly endowed with family values, who reaped handsome profits on the royalties of their best-selling textbooks.

The next generation of mathematicians was eager to create a research atmosphere, and gradually a few graduate students began to drift into New Haven. From the beginning of the graduate school until the twenties, the one notable research mathematician to have been through Yale was E. H. Moore. Prior to the fifties, only two truly distinguished mathematicians emerged: Marshall Hall and Irving Segal. But in the fifties, a sudden burst of stars appeared, led by Jack Schwartz.

It is not clear how functional analysis took over the mathematics department. Einar Hille was hired away from Princeton sometime in the mid-thirties, but for several years he was one of two research mathematicians. It was university policy to hire one mathematician as the one and only "research mathematician"; Yale could afford two: Einar Hille and Oystein Ore.

Nelson Dunford was the next arrival, as an assistant professor. Shortly thereafter, he received an attractive offer from the University of Wisconsin, whereupon Yale took the unusual step of promoting him to a full professorship.

After the end of World War II, Shizuo Kakutani came from Japan, and Charles Rickart from Michigan. By the early fifties, just about every younger mathematician at Yale was working in functional analysis, and the weekly seminars were attended by well over fifty people.

The core of graduate education in mathematics was Dunford's course in linear operators. Everyone who was interested in mathematics at Yale eventually went through the experience, even such brilliant undergraduates as Andy Gleason, McGeorge Bundy and Murray Gell-Mann. The course was taught in the style of R. L. Moore: mimeographed sheets containing unproved statements were handed out every once in a while, and the students would be asked to produce proofs on request. Once in a while some student at the blackboard fell silent, and the silence sometimes lasted an unbearable fifty minutes, since Dunford made no effort to help. Given the normal teaching demands of 12 hours a week for full professors, I suspect that he wanted to minimize his load.

Everyone who took Dunford's course was marked by it. George Seligman told me that Dunford's course in linear operators was the turning point in his graduate career as an algebraist.

Dunford had an unusual youth. After being passed over for a graduate fellowship in the middle of the depression in the thirties, he survived in St. Louis on ten dollars a month while studying and writing in the public library. Remarkably, the public library subscribed to the few mathematics journals of the time so that he managed to finish his first paper[3] dealing with integration of functions with values in a Banach space. After the paper had been accepted for publication in the *Transactions of the A.M.S.*, Dunford was offered an assistantship at Brown, working under Tamarkin. His doctoral dissertation dealt with the functional calculus that bears his name. After receiving his Ph.D. he was hired by Yale. He spent his entire career there, retiring early, ostensibly because he had made lucrative investments in art and the stock market. In reality, Dunford's retirement coincided with the completion of his life work, which was the three volume treatise *Linear Operators*,[4] written in collaboration with his student Jack Schwartz.

Linear Operators started out as a set of solutions to problems handed out in class, and gradually increased in size. Soon after Jack Schwartz enrolled in the course, Dunford asked him to become coauthor. The project, fully supported by the Office of Naval Research, quickly expanded to include Bill Bade and Bob Bartle, as well as several other students, instructors and assistant professors. There is a persistent rumor, never quite denied, that every nuclear submarine on duty carries a copy of *Linear Operators*.

Abstraction in Mathematics

The pendulum of mathematics swings back and forth towards abstraction and away from it with a timing that remains to be estimated. The period that runs roughly from the twenties to the middle seventies was an age of abstraction. It probably reached its peak in the fifties and sixties during the heyday of functional analysis and algebraic geometry, respectively.

Yale and Chicago were the two major centers of functional analysis. Stanford's mathematics department, which consisted entirely of classical analysts, had trouble finding graduate students. Pólya, Szegö, Löwner, Bergman, Schiffer and the first Spencer were great classical analysts but were considered hopelessly old-fashioned.

At Yale no courses were offered other than functional analysis and supporting abstractions. Algebra reached an independent peak of abstraction with Nathan Jacobson and Oystein Ore. There was a standing bet among Yale graduate students that if their doctoral dissertations were in analysis, they would have to use their results to give a new proof of the spectral theorem.

In those days no one doubted that the more abstract the mathematics, the better it would be. A distinguished mathematician, who is still alive, pointedly remarked to me in 1955 that any existence theorem for partial differential equations which had been proved without using a topological fixed point theorem should be dismissed as applied mathematics. Another equally distinguished mathematician once whispered to me in 1956, "Did you know that your algebra teacher Oystein Ore has published papers in graph theory? Don't let this get around!"

Sometime in the early eighties the tables turned, and a stampede away from abstraction started, which is still going on. A couple of years ago I listened to a lecture by a well known probabilist which dealt with properties of Markov processes. After the lecture, I remarked to the speaker that his presentation could be considerably shortened if he expressed his results in terms of positive operators rather than in terms of kernels. "I know," he answered, "but if I had lectured on positive operators nobody would have paid any attention!"

There are already signs that the tables may be turning again, and we old abstractionists are waiting with mischievous glee for the pendulum to swing back. Just a few months ago I overheard a conversation between two brilliant assistant professors purporting to provide an extraordinary simplification of some recently proved theorem; eventually, I realized with pleasant surprise that they were rediscovering the usefulness of taking adjoints of operators.

Linear Operators: The Past

The three volume treatise *Linear Operators* was originally meant by Dunford as a brief introduction to the new functional analysis, and to the spectral theory which had been initiated by Hilbert and Hellinger but had not taken root until the work of von Neumann and Stone. It was Dunford, however, who championed spectral theory as a new field. He introduced the term "resolution of the identity," and he developed the program of extending spectral theory to non-selfadjoint operators.

The initial core of the book consisted of what are now Chapters Two, Four, and Seven, as well as some material on spectral theory now in Chapter Ten; eventually this material expanded into two volumes. The idea of volume three was a belated one, coming in the wake of the development of the theory of spectral operators.

The writing of *Linear Operators* took approximately twenty years, starting in the late forties. The third volume was published in 1971. Entire sections and even entire chapters were added to the text at various times up to the last minute. For example, one of the last bits to be added to the first volume right before it went to press was the last part of Section Sixteen of Chapter Four, containing the Gauss-Wiener integral in Hilbert space together with a simple formula relating it to the ordinary Wiener integral. This section was the subject of a lecture that Jack Schwartz gave at the Courant Institute in 1956 at the famous seminar on integration in function space.

The flavor of the first drafts of the book can be gleaned from reading Chapter Two, which underwent fewer drafts than most of the other chapters. Dunford meant the three theorems proved in his chapters, namely the Hahn-Banach Theorem, the Uniform Boundedness Theorem and the Closed Graph Theorem, to be the cornerstones of functional analysis. The exercises for this chapter, initially rather dry, were eventually enriched by a set of exercises on summability of series. These problems continue in Chapter Four, and conclude in Chapter Eleven with the full expanse of Tauberian theorems. The contrast between the uncompromising abstraction of the text and the incredible variety of the concrete examples in the exercises is immensely beneficial to any student learning mathematical analysis from Dunford-Schwartz.

The topics dealt with in Dunford-Schwartz can be roughly divided into three types. There are topics for which Dunford-Schwartz is still the definitive account. There are, on the other hand, other topics fully dealt with in the text which ought to be well-known, but which have yet to be properly read. Finally, there are topics still ahead of the times, which remain to be fully appreciated. Presumptuous as it is on my part, I will try to give some examples of each type.

Besides the introductory chapter on Banach spaces (Chapter Two), the treatment of the Stone-Weierstrass Theorem and all that goes with it in Chapter Four makes very enjoyable reading; it was the first thorough account. The short sections on Bohr compactification and almost periodic functions are also the best reference for a quick summary of Bohr's extensive theory.

Section twelve of Chapter Five, presenting a proof of the Brouwer fixed point theorem, is remarkable. The proof was submitted for publication in a journal in 1954, but was rejected by an irate referee, a topologist who was miffed by the fact that the proof uses no homology theory whatsoever. Instead, the proof depends on some determinantal identities, the kind that are again becoming fashionable.

Spectral theory proper does not make an appearance until Chapters Seven and Eight, with the functional calculus and the theory of semigroups. In those days, such terms as "resolvent" and "spectrum" carried an aura of mystery, and the spectral mapping theorem sounded like magic.

The meat and potatoes come in Chapters Ten, Twelve, and Thirteen; the proofs are invariably the most instructive, bringing into full play the abstract theory of boundary conditions of Calkin and von Neumann, as well as the theory of deficiency indices.

Linear Operators: The Present

There are topics for which Dunford-Schwartz was the starting point of a long development, and which have grown into autonomous subjects. For example, the notion of unconditional convergence of series in Banach spaces, which goes back to an old theorem of Steinitz and is mentioned in Chapter Two almost as a curiosity, has blossomed into a full-fledged discipline. The same can be said of the geometry of Banach spaces initiated in Chapter Four, and of the theory of convexity in Chapter Five. In the sixties, several mathematicians pronounced the general theory of Banach spaces dead several times over, but this is not what happened. The geometry of Banach spaces has not only managed to survive, but it is now widely considered to be the deepest chapter of convex geometry. Grothendieck told me that his favorite theorem of his analysis period was a convexity theorem that generalizes a result in Dunford-Schwartz. He published it in an obscure Brazilian journal, and never received any reprints of the paper.[5]

The theory of vector-valued measures in Chapter Four has also blossomed into a beautiful and deep chapter of functional analysis. At the time of the book's writing, we all thought that this theory had reached its definitive stage, perhaps because the proofs were so crystal clear.

The same can be said of the theory of representation of linear operators in Chapter Six; here again whole theories nowadays replace single sections of Dunford-Schwartz. Corollary five of Section Seven, stating that in certain circumstances the product of two weakly compact operators is a compact operator, has always struck me as one of the most elegant results in functional analysis. Undoubtedly sooner or later some extraordinary application of it will be found, as should happen to all beautiful theorems.

Thorin's proof of the Riesz convexity theorem had appeared a short time before Chapter Six was written, and is given its first billing in a textbook here. I take the liberty of calling your attention to Problem Fifteen of Section Eleven. This exercise holds the key to giving one line proofs of some of the famous inequalities in the classic book by Hardy, Littlewood, and Pólya.

Section nine of Chapter Twelve has been scandalously neglected. The classical moment problems are thoroughly dealt with in this section by an application of the spectral theorem for unbounded selfadjoint operators. It is shown in a couple of pages that the various criteria for determinacy of the moment problem can be inferred from a simple computation with deficiency indices. Partial rediscoveries of this fact are still being published every few years by mathematicians who haven't done their reading.

Linear Operators: The Future

Finally, there are a number of subjects that were first written up in Dunford-Schwartz from which the mathematical world has yet to benefit. It is surprising to hear from time to time probabilists or physicists addressing problems for which they would find help in Dunford-Schwartz. Actually, Chapter Three already contains a number of yet-to-be appreciated jewels. One of them is the comprehensive treatment of theorems of Vitali-Hahn-Saks type. The proofs are so concocted as to bring out the analogies between the combinatorics of sigma-fields of sets and the algebra of linear spaces. Few analysts still make use of this kind of reasoning. In probability, an appeal to the Vitali-Hahn-Saks theorem would bypass technical complications that are instead settled by the Choquet theory: for example, randomization theorems of De Finetti type. Apparently the only probabilist to have taken advantage of this opportunity is Alfred Rényi, in an elementary introduction to probability that also has been little read. Similarly, one wonders why so little use is made of Theorem III.7.8, which might come in handy in integral geometry.

Large portions of spectral theory presented in Dunford-Schwartz remain to be assimilated. The fine theory of Hilbert-Schmidt operators and the wholly original theory of subdiagonalization of compact operators in Chapter Eleven have not been read. The spectral theory of non-selfadjoint operators of Chapters Fifteen through Nineteen is an untapped gold mine. Only the latter parts of Chapter Twenty, dealing with what the authors have successfully called "Friedrichs' method" and with the wave operator method, have been developed since the treatise was published. It will be a pleasure to watch the rediscovery of these chapters by younger generations.

Working with Jack Schwartz

There are fringe benefits to being a student of Jack Schwartz. I decline invitations to attend meetings in computer science and even economics from organizers who mistakenly assume that I have inherited my thesis advisor's interests.

Two traits have particularly endeared Jack to his students: the first is his instinctive understanding of another person's state of mind and his tact in dealing with difficult situations. He gives encouragement without exaggerating, and he knows how to steer his friends away from being their own worst enemies.

The second is his Leibnizian universality. This quality seems to spill over onto all of us, lifting and pointing us in the right direction. Whatever topic he deals with at one point is envisioned as a stepping stone to a wider horizon to be dealt with in the future.

Both of these qualities shine in the pages of *Linear Operators*, the first by the transparent proofs, the second by the encyclopedic range of the material in the 2592 pages.

I was hired together with Dave McGarvey to work on the Dunford-Schwartz project in the summer of 1955. Immediately Jack took us aside and let us in on the delicate matter of the semicolons. There were to be no semicolons in anything we wrote for the project. Dunford would get red in the face every time he saw one. For years thereafter I was terrified of being caught using a semicolon, and you may verify that in the three printed volumes of Dunford-Schwartz not a single semicolon is to be found.

I was asked to check the problems in Chapter Three, while Dave was checking problems in Chapters Seven and Eight. We would all get together every morning in a little office in Leet Oliver Hall, an office that nowadays would not be considered fit for a teaching assistant. A bulky record player, which we had bought for ten dollars, occupied much of the space; we played over and over the entire sequence of Beethoven quartets and Bach partitas while working on the problems.

It took me half the summer to finish checking the problems in Chapter Three. There were a few that I had trouble with, and worst of all, I was unable to work out Problem Twenty of Section Nine. One evening Dunford and several other members of the group got together to discuss changes in the exercises. Jack was in New York City. It was a warm summer evening and we sat on the hard wooden chairs of the corner office of Leet Oliver Hall. Pleasant sounds of squawking crickets and frogs along with mosquitoes came through the open gothic windows. After I admitted my failure to work out Problem Twenty, Dunford tried one trick after another on the blackboard in an effort to solve the problem or to find a counterexample. No one remembered where the problem came from, or who had inserted it.

After a few hours, feeling somewhat downcast, we all got up and left. The next morning I met Jack, who patted me on the back and told me, "Don't worry, I could not do it either." I did not hear about Problem Twenty of Section Nine for another three years. A first-year graduate student had taken Dunford's course in linear operators. Dunford had assigned him the problem, the student solved it, and developed an elegant theory around it. His name is Robert Langlands.

In the second half of the summer of 1955, after checking the problems in Chapter Three, I was assigned to check the problems in spectral theory of differential operators in Chapter Thirteen. This is the chapter of Dunford-Schwartz that decided my career in mathematics. I had less difficulty with this second round of exercises, but made a number of careless mistakes.

One day I was unexpectedly called in by Dunford. The details of this meeting have been rewritten in my mind many times. The large office was empty, except for Dunford and Schwartz sitting together at the desk in the shadows, like judges.

"We have decided to assign you the problems in Sections G and H of Chapter Thirteen," they said. A minute of silence followed. I had the feeling that there was something they were not saying. Eventually I got it. They were NOT assigning me the problems in Section I which dealt with the use of special functions in eigenfunction expansions. I soon learned, somewhat to my annoyance, that the person in charge of checking those problems was an undergraduate who had just gotten his B.A. two months earlier. "You will never find a better undergraduate in math coming out of Yale," Jack told me, aware of my feelings. He was right. The undergraduate checked all the special function problems by the end of the summer, and Section I is now spotless. His name is John Thompson. He went on to win the Fields Medal.

I have kept a copy of the mimeographed version of the manuscript of Dunford-Schwartz. On gloomy days I pull the dusty fifteen-pound bulk off the shelf. Reading the yellowed pages with their inky smell was once a great adventure; rereading them after forty years is a happy homecoming.

CHAPTER III

Combinatorics, Representation Theory and Invariant Theory

The Story of a Ménage à Trois

To Adriano Garsia on his sixty-fifth birthday

We will dwell upon the one topic of unquestioned interest and timeliness among mathematicians: gossip. Or rather, to use an acceptable euphemism, we will deal with the *history* of mathematics.

Cambridge 02138 in the Early Fifties

The fifties were a great time to be alive, and the assembly of younger mathematicians who went through the universities in the Boston area would now fill the *Who's Who of Mathematics*.

The center of mathematical activity was the MIT common room, renovated in 1957, which has been allowed to deteriorate ever since. At frequent intervals during the day, one could find in the MIT common room Paul Cohen, Eli Stein, and later Gene Rodemich, excitedly engaged in aggressive problem solving sessions and other mutual challenges to their mathematical knowledge and competence.

The leader of the problem solving sessions was without question Adriano Garsia, at times joined by cameo appearances of Alberto Calderón, Jürgen Moser, and John Nash. Often a discussion that had started in the common room would be carried over uninterrupted through lunch to Walker, where mathematicians used to assemble around a large table, easy to spot amid the hustle and bustle of faculty and students.

Norbert Wiener would often join the younger mathematicians at lunch. He loved to sit at the old wooden tables in Walker. He glowed under the stares of the undergraduates and craved the fawning admiration of the younger mathematicians. The temptation to tease him was irresistible.

One day several of us were having lunch at the usual table in Walker. Norbert Wiener sat at the head, with Paul Cohen at his right; others at the table were probably Adriano Garsia, Arthur Mattuck, myself, and some other person whose name I cannot recall. Cohen turned towards Wiener and asked in a tone of mock candor, "Professor Wiener, what would you do if one day when you went home, you were to find Professor X sitting on your living room sofa?" Cohen was alluding to a well-known mathematician who was known to indulge in the dubious practice of "nostrification." Wiener became red in the face and snapped back, "I would throw him out and start counting the silver!" I leave it to you to figure out who Professor X is. By the way, the term "nostrification" was introduced by Hilbert, and the practice has been faithfully carried on by his students.

From time to time, the problem sessions in the MIT common room were temporarily suspended and replaced by "ranking sessions," in which all of us instructors would indulge in the favorite hobby of younger mathematicians, that of passing judgment on older mathematicians and listing them in strict linear order. I remember a heated discussion about whether Professor Y should be rated as a first-rate second-rate mathematician or a second-rate first-rate mathematician.

Every time I prepared to enter the MIT common room, I had to be sure to swallow an extra tablet of Nodoz. Eventually I began to avoid the common room altogether. I could not take the heat and Adriano Garsia agreed to meet me privately to bring me up to date on the latest happenings. All the differential geometry I know I learned from these sessions with him.

One month in the spring of 1958 he was kind enough to find the time for a series of ten lectures on the theory of certain surfaces, a theory he had developed in his first year at MIT. He had named them "Schottky surfaces" to honor Schottky, an otherwise obscure mathematician (or so we believed at the time). I never heard Schottky's name mentioned again until Paul Erdös told me a startling story, which you will read shortly.

Neither Adriano nor I had any inkling that we would end up working in combinatorics. The term "combinatorics" was all but unknown. The problems of the day, those by which mathematicians test one another, were frequently drawn from analysis (such as one finds in Pólya and Szegö's collection), rather than from combinatorics. Adriano was fond of reminiscing about two of his teachers, both of whom came from the same great German school of analysts as Pólya and Szegö: Karl Löwner and Marcel Riesz. After I met Löwner and Riesz I noticed a family resemblance.

At first I did not believe some of the stories Adriano used to tell me about Marcel Riesz: for example, the story about Riesz hiding his paychecks under the mattress instead of cashing them, and Adriano having to run to banks all over town cashing the checks the day before Riesz was scheduled to leave for Sweden. Later, I came to realize these stories were true.

Alfred Young

Alfred Young believed his greatest contribution to mathematics was the application of representation theory to the computation of invariants of binary forms. If he had been told that one day we would mention his name with reverence in connection with the notion of standard tableaux, he probably would have winced.

The story of standard tableaux is an interesting episode in mathematical history. Alfred Young made his debut in mathematics with the difficult computation of the concomitants of binary quartics, a *tour de force* which followed from Peano's elegant and unjustly forgotten finiteness theorem.

As Young proceeded to derive a systematic method for computing the syzygies holding among the invariants of such quartics, he realized that the methods developed by Clebsch and Gordan could not be pushed much farther. He went into a period of self-searching which lasted a few years after which he published the first two papers in the series "Quantitative Substitutional Analysis."[1] Both papers appeared in short sequence at the turn of the century. In them Young outlined the theory of representations of the symmetric group. He proved that the number of irreducible representations of the symmetric group of order n equals the number of partitions of n, and he gave an explicit decomposition of the group algebra into irreducible components by means of idempotents.

Young's combinatorial construction of the irreducible representations of the symmetric group remains the simplest though not the most elegant. He made no appeal to the theory of group representations (or group characters, as it was called by Frobenius who had developed it). The term "group" seldom appears in the seven hundred-odd pages of his collected papers. It is safe to surmise that Young was reluctant to rely upon group-theoretic arguments; for example, he never used the expression "normal subgroup."

It seems that Young's results irritated the leading algebraist of the time, Professor Frobenius of the University of Berlin. What? This British upstart, using rudimentary combinatorial methods, could get what everyone in his sophisticated German school had been working on for years? Frobenius carefully studied Young's papers (while skipping Young's elaborate applications to invariant theory, a subject which Frobenius despised) and went to work. After a short time, he published a paper in which Young's results were properly rederived following the precepts of his own newly invented theory of group characters. Frobenius was a Bourbakist at heart, and his way of doing representation theory prevails to this day. He went one up on Alfred Young by discovering the character formula that now bears his name, a formula which Young had unforgiveably missed.

Young was deeply hurt when he learned of Frobenius' work. Worse yet, at exactly the same time as he was getting ready to publish his two papers on substitutional analysis, Frobenius assigned to his best student Issai Schur the thesis problem of determining all possible generalizations of the Binet-Cauchy formula for the multiplication of minors of matrices. In 1900 Issai Schur published his thesis in which all irreducible representations of the general linear group were explicitly determined on the basis of their traces, now called Schur functions. Young realized the connection between his work and Schur's thesis. Although they did not overlap, Young's two papers and Schur's thesis contained embarrassingly close results.

For some twenty years after the turn of the century, Young did not publish another paper. To surprised colleagues making tactful inquiries, he would answer that he was learning to read German in order to understand Frobenius' work. But this was a white lie. Young was intensively working on going one up on Frobenius, and he did. In 1923, some twenty years after his second paper, he published his third paper on quantitative substitutional analysis. In this paper standard tableaux were first introduced, their number was computed, and their relation to representation theory was described. Again Young's methods were purely combinatorial, containing not one iota of group theory or character theory. A new proof of Frobenius' character formula was given. Disregarding the apparatus of representation theory, Young used purely combinatorial techniques.

Again a period of silence ensued. It appears that the German algebraists were lending a deaf ear to Young's discovery. Had it not been for Hermann Weyl's intervention, Young's newly discovered standard tableaux might have been permanently exiled to Aberystwyth, Wales, together with apolarity and perpetuants.

Hermann Weyl, while writing his book on group theory and quantum mechanics in the twenties, came upon Young's work and realized its importance. The term "Young tableau" first appeared in print in Hermann Weyl's book *Group Theory and Quantum Mechanics*[2] (a book which was rewritten and updated some forty years later by Leona Schensted). Since Hermann Weyl represented the pinnacle of mainstream mathematics, it did not take long for Young's name to become a household word, first with physicists and then mathematicians. Van der Waerden included standard tableaux in one of the later editions of his *Modern Algebra*[3] in which he made use of some unpublished proofs of von Neumann. The geometer Hodge learned about standard tableaux from Young's associate, D. E. Littlewood, and used them with success in his study of flag manifolds. The algebras that are named after Hodge might more justly have been named after Young. Hodge acknowledges his indebtedness to Littlewood and Young.

Alfred Young was generous with his ideas. Turnbull used to come down to Cambridge (the *other* Cambridge) from St. Andrews once a month to talk to Young. Young's ideas on the future of invariant theory were made public in Aitken's *Letter to an Edinburgh Colleague*. The various equivalent definitions of Schur functions first appeared together in Philip Hall's paper "The Algebra of Partitions" without any proofs; for several years after Hall's paper, no printed proofs of the equivalence of these definitions were available, and we had to construct our own.

Alfred Young's style of mathematical writing has unfortunately gone out of fashion: it is based on the assumption that the reader is to be treated as a gentleman with a sound mathematical education, and gentlemen need not be told the lowly details of proofs. As a consequence, we have to figure out certain inferences for which Young omits any explanation out of respect for his readers.

Young's one and only student G. de B. Robinson wrote the ninth and last paper on quantitative substitutional analysis on the basis of notes left by his teacher. Robinson inherited all of Young's handwritten papers after his teacher's death. I will not conjecture about the fate of these papers. Suffice it to say that, some fifteen years ago, Alfred Young's collected papers were published by the Toronto University Press at the price of ten dollars a copy, a prize subsidized by an unknown donor. They are still in print.

Problem Solvers and Theorizers

Mathematicians can be subdivided into two types: problem solvers and theorizers. Most mathematicians are a mixture of the two although it is easy to find extreme examples of both types.

To the problem solver, the supreme achievement in mathematics is the solution to a problem that had been given up as hopeless. It matters little that the solution may be clumsy; all that counts is that it should be the first and that the proof be correct. Once the problem solver finds the solution, he will permanently lose interest in it, and will listen to new and simplified proofs with an air of condescension suffused with boredom.

The problem solver is a conservative at heart. For him, mathematics consists of a sequence of challenges to be met, an obstacle course of problems. The mathematical concepts required to state mathematical problems are tacitly assumed to be eternal and immutable.

Mathematical exposition is regarded as an inferior undertaking. New theories are viewed with deep suspicion, as intruders who must prove their worth by posing challenging problems before they can gain attention. The problem solver resents generalizations, especially those that may succeed in trivializing the solution of one of his problems.

The problem solver is the role model for budding young mathematicians. When we describe to the public the conquests of mathematics, our shining heroes are the problem solvers.

To the theorizer, the supreme achievement of mathematics is a theory that sheds sudden light on some incomprehensible phenomenon. Success in mathematics does not lie in solving problems but in their trivialization. The moment of glory comes with the discovery of a new theory that does not solve any of the old problems but renders them irrelevant.

The theorizer is a revolutionary at heart. Mathematical concepts received from the past are regarded as imperfect instances of more general ones yet to be discovered. Mathematical exposition is considered a more difficult undertaking than mathematical research.

To the theorizer, the only mathematics that will survive are the definitions. Great definitions are what mathematics contributes to the world. Theorems are tolerated as a necessary evil since they play a supporting role — or rather, as the theorizer will reluctantly admit, an essential role — in the understanding of definitions.

Theorizers often have trouble being recognized by the community of mathematicians. Their consolation is the certainty, which may or may not be borne out by history, that their theories will survive long after the problems of the day have been forgotten.

If I were a space engineer looking for a mathematician to help me send a rocket into space, I would choose a problem solver. But if I were looking for a mathematician to give a good education to my child, I would unhesitatingly prefer a theorizer.

Hermann Grassmann and Exterior Algebra

Alfred Young was more of a problem solver than a theorizer. But one of the great mathematicians of the nineteenth century was a theorizer all the way: I mean Hermann Grassmann. Everyone agrees that Grassmann's one great contribution to mathematics is a new definition, namely the definition of exterior algebra. He gave his entire life to understanding and developing this definition.

Grassmann never solved any of the problems that were fashionable in his day; he never solved any problems at all, except those which he himself had posed. He never contributed to the mathematics of the nineteenth century: invariant theory, elimination theory, the theory of algebraic curves, or any of the current fads. What is worse, to the dismay of his contemporaries, he rewrote some of the mathematics of his time in the language of exterior algebra. He was the first to show that much classical physics could be simplified in the notation of exterior algebra, thereby anticipating the calculus of exterior differential forms developed by Elie Cartan. The work of Gibbs and then Dirac would have been considerably simplified had Gibbs and Dirac even a fleeting acquaintance with exterior algebra. Instead, exterior algebra ended up as another missed opportunity, as Freeman Dyson might say.

It is not surprising that Grassmann was not entirely welcome among mathematicians. Anyone coming up with a new definition is likely to make enemies. New ideas are often unwelcome and regarded as intrusive. Grassmann made a number of enemies, and the animosity against his great definition has not entirely died out.

The reactions against Grassmann make a humorous chapter in the history of mathematics. For example, Professor Pringsheim, dean of German mathematicians and the author of over one hundred substantial papers on the theory of infinite series, both convergent *and* divergent, kept insisting that Grassmann should be doing something relevant instead of writing up his maniacal ravings. "Why doesn't he do something useful, like discovering some new criterion for the convergence of infinite series!" Pringsheim asserted, with all the authority that his position conferred.

The invariant theory community led by Clebsch and Gordan also loudly protested Grassmann's work as pointless since it did not contribute one single result to the invariant theory of binary forms. They were dead wrong, but would not be proved so for another fifty years.

Not even Hilbert paid attention to Grassmann. In the second volume of his collected papers I found only one mention of Grassmann, in a footnote. And Eduard Study, editor of Grassmann's collected papers,[4] only partially understood exterior algebra. Study's last book, *Vector Algebra*[5] (in German, *Einleitung in die Theorie der Invarianten Lineärer Transformationen auf Grund der Vektorrechnung*) would have greatly benefited from an injection of exterior algebra. It is clear, however, that Study did not feel comfortable enough with exterior algebra to make use of it in his work.

Evil tongues whispered that there was really nothing new in Grassmann's exterior algebra, that it was just a mixture of Möbius' barycentric calculus, Plücker's coordinates, and von Staudt's algebra of throws. The standard objection was expressed by the notorious question, "What can you prove with exterior algebra that you cannot prove without it?" Whenever you hear this question raised about some new piece of mathematics, be assured that you are likely to be in the presence of something important. In my time, I have heard it repeated for random variables, Laurent Schwartz' theory of distributions, idèles and Grothendieck's schemes, to mention only a few. A proper retort might be: "You are right. There is nothing in yesterday's mathematics that you can prove with exterior algebra that could not also be proved without it. Exterior algebra is not meant to prove old facts, it is meant to disclose a new world. Disclosing new worlds is as worthwhile a mathematical enterprise as proving old conjectures."

The first mathematician to understand the importance of exterior algebra was Peano who published a beautiful short introduction to the subject.[6] At the time Peano was teaching at the Pinerolo military school. His audiences for what must have been beautiful lectures on exterior algebra consisted of Italian cavalry officers and cadets. No one living beyond the Alps read Peano's book until Elie Cartan came along. Three hundred copies were printed of the first and only edition.

It took almost one hundred years before mathematicians realized the greatness of Grassmann's discovery. Such is the fate meted out to mathematicians who make their living on definitions.

Definition and Description in Mathematics

One of the great achievements of mathematics in this century is the idea of precise definitions ensconced in an axiomatic system. A mathematical object must and can be precisely defined; this is the only way we have to make sure we are not dealing with pure fantasy.

While stressing the importance of definition, our century has given short shrift to an older notion, that of the description of a mathematical object. Description and definition are two quite different enterprises, sometimes confused with each other. The difference between definition and description can be understood by performing the following thought experiment. Suppose you are trying to teach a new mathematical concept to your class. You know that you cannot get away with just writing a definition on the blackboard. Sooner or later you must describe what is being defined. Nowadays one of the more common ways of describing a new mathematical object is to give several equivalent definitions of it. Philosophers have long puzzled over this strange phenomenon, whereby completely different definitions can be given of the same mathematical object.

In ages past, mathematical objects were described before they could be properly defined, except in geometry, where Euclid set the standards early in the game, and the need for a precise definition was not even felt. The mathematics of the past two centuries confirms the fact that mathematics can get by without definitions but not without descriptions. Physicists have long been aware of this priority.

To be sure, mathematicians of all times claim that definition is a *sine qua non* of a proper mathematical presentation. But mathematicians, like other people, seldom practice what they preach.

Take the field of real numbers, which was not defined by current standards of rigor until Dedekind came along very late in the game. Shall we conclude that, given the lack of a rigorous definition, all work on the real numbers before Dedekind's time is to be discarded? Certainly not.

Another equally glaring example is the concept of a tensor. When I was an undergraduate at Princeton, Professor D. C. Spencer defined a tensor as "an object that transforms according to the following rules"; this is the description of a tensor that you will also find in Luther Pfahler Eisenhart's textbook on differential geometry,[7] still considered the best introductory textbook on the subject. It was clear to everyone that such a nonsensical statement was not a definition. Every time such a characterization of a tensor was stated, it was followed by a slight giggle. Nevertheless, the lack of definition of a tensor did not stop Einstein, Levi-Civita and Cartan from doing some of the best mathematics in this century.

As a matter of fact, the first correct definition of a tensor did not become current until the fifties, under the influence of Chevalley I believe, or perhaps I should say Bourbaki. Even more amazing, the first completely rigorous definition of a tensor was given just at the time when tensors were going out of fashion—temporarily.

A lot of mathematical research time is spent in finding suitable definitions to justify statements that we already know to be true. The most famous instance is the Euler-Schläfli-Poincaré formula for polyhedra, which was believed to be true long before a suitably general notion of a polyhedron could be defined.

At least one hundred years of research were spent on singling out a definition to match the Euler-Schläfli-Poincaré formula. Meanwhile, no one ever entertained any doubt of the formula's truth. The philosopher Imre Lakatos has documented the story of such a search for a definition in thorough historical detail. Curiously, his findings, published in the book *Proofs and Refutations*,[8] were met with a great deal of anger on the part of the mathematical public who held the axiomatic method to be sacred and inviolable. Lakatos' book became anathema among philosophers of mathematics of the positivistic school. The truth hurts.

Hermann Grassmann was a great believer in description. By and large he did not bother to give definitions in the current sense of the term. His descriptive style, coming precisely at the time when the axiomatic method was becoming a fanatical devotion among mathematicians, is very probably one more reason why his work was not read. As a matter of fact, the first rigorous definition of exterior algebra was not given until the forties, by Bourbaki in Chapter Three of *Algèbre*,[9] which is perhaps the best written of all Bourbaki's volumes. Each successive edition of this chapter *"fait regretter les précédentes."*

Every mathematician of my generation and the preceding learned exterior algebra from Bourbaki. Emil Artin, for example, did so in the early fifties, motivated by the Galois cohomology he was inventing at the time.

I cannot refrain from telling you a story about myself. Sometime in 1951 I traveled from Princeton to New York to visit Stechert-Hafner on Fourth Avenue, a huge four-story academic bookstore that has since gone bankrupt. Stechert-Hafner looked more like a warehouse than a bookstore; books were spread all over in no particular order, ready to be shipped to some college library. As I came out of the elevator on the third floor, I walked up to a lady who was working with an adding machine and seemed to be the only person present. After a few minutes she glanced in my direction. She looked at me squarely and before I could speak a word she said, "I know your type! You want the Bourbáki books!" She was right. She gave me a huge discount and I will never forget her.

Bottom Lines

How do mathematicians get to know each other? Professional psychologists do not seem to have studied this question; I will try out an amateur theory. When two mathematicians meet and feel out each other's knowledge of mathematics, what they are really doing is finding out what each other's bottom line is. It might be interesting to give a precise definition of a bottom line; in the absence of a definition, we will give some typical examples.

To the algebraic geometers of the sixties, the bottom line was the proof of the Weil conjectures. To generations of German algebraists, from Dirichlet to Hecke and Emil Artin, the bottom line was the theory of algebraic numbers. To the Princeton topologists of the fifties, sixties, and seventies, the bottom line was homotopy. To the functional analysts of Yale and Chicago, the bottom line was the spectrum. To combinatorialists, the bottom lines are the Yang-Baxter equation, the representation theory of the classical groups, and the Schensted algorithm. To some algebraists and combinatorialists of the next ten or so years, the bottom line may be elimination theory.

I will shamelessly tell you what my bottom line is. It is placing balls into boxes, or, as Florence Nightingale David put it with exquisite tact in her book *Combinatorial Chance*,[10] it is the theory of distribution and occupancy.

We resort to the bottom line when we are asked to write a letter of support for some colleague. If the other mathematician's bottom line is agreeable with our own, then our letter is more likely to be positive. If our bottom lines disagree, then our letter is likely to be restrained.

The most striking example of mismatch of bottom lines was told to me by Erdös. When David Hilbert, then a professor at the University of Königsberg, was being considered for a professorship at Göttingen, the Prussian ministry asked Professor Frobenius to write a letter in support of Hilbert's candidacy. Here is what Frobenius wrote: "He is rather a good mathematician, but he will never be as good as Schottky."

Allow me to tell you two more personal stories. In 1957, in my first year as an instructor in Cambridge, I often had lunch with Oscar Zariski, who liked to practice his Italian. One day while we were sitting in the main room of the Harvard Faculty Club he peered at me, fork in hand and said, loudly enough for everyone to hear, "Remember! Whatever happens in mathematics happens in algebraic geometry first!" Algebraic geometry has been the bottom line of mathematics for almost one hundred years; but perhaps times are changing.

The second story is more somber. One day, in my first year as an assistant professor at MIT, while walking down one of the long corridors, I met Professor Z, a respected senior mathematician with a solid international reputation. He stared at me and shouted, "Admit it! All lattice theory is trivial!" I did not have the presence to answer that von Neumann's work in lattice theory is deeper than anything Professor Z has done in mathematics.

Those who have reached a certain age remember the visceral and widespread hatred of lattice theory from around 1940 to 1979; this has not completely disappeared. Such an intense and unusual disliking for an entire field cannot be simply attributed to personality clashes. It is more likely to be explained by pinpointing certain abysmal differences among the bottom lines of the mathematicians of the time. If we begin such a search, we are likely to conclude that the field normally classified as algebra really consists of two quite separate fields. Let us call them Algebra One and Algebra Two for want of a better language.

Algebra One is the algebra whose bottom lines are algebraic geometry or algebraic number theory. Algebra One has by far a better pedigree than Algebra Two, and has reached a high degree of sophistication and breadth. Commutative algebra, homological algebra, and the more recent speculations with categories and topoi are exquisite products of Algebra One. It is not infrequent to meet two specialists in Algebra One who cannot talk to each other since the subject is so vast. Despite repeated and dire predictions of its demise, Algebra One keeps going strong.

Algebra Two has had a more accidented history. It can be traced back to George Boole, who was the initiator of three well-known branches of Algebra Two: Boolean algebra, the operational calculus that views the derivative as the operator *D*, on which Boole wrote two books of great beauty, and finally, invariant theory, which Boole initiated by remarking the invariance of the discriminant of a quadratic form.

Roughly speaking, between 1850 and 1950 Algebra Two was preferred by the British and the Italians, whereas Algebra One was once a German and lately a French preserve. Capelli and Young's bottom lines were firmly in Algebra Two, whereas Kronecker, Hecke, and Emil Artin are champions of Algebra One.

In the beginning Algebra Two was largely cultivated by invariant theorists. Their objective was to develop a notation suitable to describe geometric phenomena which is independent of the choice of a coordinate system. In pursuing this objective, the invariant theorists of the nineteenth century were led to develop explicit algorithms and combinatorial methods. The first combinatorialists, MacMahon, Hammond, Brioschi, Trudi, Sylvester, were invariant theorists. One of the first papers in graph theory, in which the Petersen graph is introduced, was motivated by a problem in invariant theory. Clifford's ideal for invariant theory was to reduce the computation of invariants to the theory of graphs.

The best known representative of Algebra Two in the nineteenth century is Paul Gordan. He was a German, perhaps the exception that tests our rule. He contributed a constructive proof of the finite generation of the ring of invariants of binary forms which has never been improved upon, and which foreshadows current techniques of Hopf algebra. He also published in 1870 the fundamental results of linear programming, a discovery for which he has never been given proper credit. Despite his achievements, Paul Gordan was never fully accepted by specialists in Algebra One. "*Er war ein Algorithmiker*!" said Hilbert when Gordan died.

Gordan's student Emmy Noether became an ardent apostle of Algebra One; similarly, van der Waerden, a student of General Weitzenböck, an Algebra Two hero, intensely disliked Algebra Two throughout his career. In the thirties, Algebra Two was enriched by lattice theory and by the universal algebra of Philip Hall and his student Garrett Birkhoff.

Algebra Two has always had a harder time. You won't find lattices, exterior algebra, or even a mention of tensors in any of the editions of van der Waerden's *Modern Algebra*. G. H. Hardy subtly condemned Algebra Two in England in the latter half of the nineteenth century with the exclamation, "Too much $f(D)$!" G. H. Hardy must be turning in his grave now.

Algebra Two has recently come of age. In the last twenty years or so, it has blossomed and acquired a name of its own: algebraic combinatorics. Algebraic combinatorics, after a tortuous history, has at last found its own bottom line, together with a firm place in the mathematics of our time.

CHAPTER IV

The Barrier of Meaning

It took me several years to realize what Stan Ulam's real profession was. Many of us at the Los Alamos Laboratory who were associated with him knew how much he disliked being alone, how he would summon us at odd times to be rescued from the loneliness of some hotel room, or from the four walls of his office after he had exhausted his daily round of long distance calls.

One day I mustered the courage to ask him why he constantly wanted company, and his answer gave him away. "When I am alone," he admitted, "I am forced to think things out, and I see so much that I would rather not think about!" I then saw the man in his true light. The man who had the highest record of accurate guesses in mathematics, the man who could beat engineers at their game, who could size up characters and events in a flash, was a member of an all-but-extinct profession, the profession of prophet.

He shared with the men of the Old Testament and with the Oracle at Delphi the heavy burden of instant vision. And like all professional prophets he suffered from a condition that Sigmund Freud would have labeled a "Proteus complex." It is unfortunate that Freud did not count any prophets among his patients.

In bygone days the Sybil's dark mumblings would have been interpreted by trained specialists, hermeneuts they were called, who were charged with the task of expressing her cryptic messages in lapidary Greek sentences. In Ulam's case, the Los Alamos Laboratory would instead hire consultants charged with expressing his cryptic messages in the dilapidated jargon of modern mathematics.

Luckily, some of Stan Ulam's pronouncements were tape-recorded by Françoise Ulam. When they are edited and published, the younger generations will find, if they are still willing to read, that some ideas they believe to be their own had been anticipated in his flashes of foresight.

I wish to report on one such conversation. It was one of the longest I ever had with him on a single subject; it lasted perhaps as long as fifteen minutes. We were strolling along the streets of Santa Fe high up on a hill. It was a winter day and the barren Latin cityscape contrasted with the snow melting under the shining sun. It was the time when everyone was discussing the A.I. problem, the problem of giving a description of intelligent behavior that would be precise enough to make computer simulation possible.

"Everyone nowadays seems to be concerned with this problem," I said. "Neurophysiologists and psychologists who work on the problem of perception are joining forces with computer scientists, and even the philosophers are hoping against hope to be taken seriously, and thereby to get a salary raise."

"Yes," continued Stan, "but they will all find out that what they now call the A.I. problem has been around for a long time. Philosophers and logicians, from Duns Scotus in the twelfth century to Ludwig Wittgenstein only yesterday, have done a lot of clear thinking about all this stuff. If your friends in A.I. persist in ignoring their past, they will be condemned to repeat it, at a high cost that will be borne by the taxpayers."

"I think I know what you mean," I added, eager to put in my two bits of wisdom. "The basic problem of A.I. boils down to the old puzzle of infinite regression. When you look at an object, that tree out there say, how is the tree registered in your brain? All explanations ever proposed pin down this process on some device inside the brain that is in charge of doing the registering. This device is supposed to behave like a little man inside the head, like in a Gary Larson cartoon. But then the same problem has to be faced again in order to explain how the little man does the registering, and so on, *ad infinitum*."

"Very good," said Stan, patting me on the back. "And I don't have to tell you that René Descartes himself worried about this problem for a long time. In the end, he concluded that the only possible explanation was Divine intervention. Quite a shocking conclusion coming from the man who founded modern rationalism."

"Too bad we cannot build thinking machines with the help of Divine intervention," I retorted, "at least not with His *direct* help. But you must admit that A.I. has the merit of focusing on the practical aspects of this question. It has the option of performing experiments with big computers, a luxury that was denied to Descartes who had to live exclusively by his wits."

"Certainly," said Stan, "but even in the age of big computers someone still has to do the thinking, and Descartes is going to be a hard act to follow."

"What then do you propose we think about?" I asked, with ill-concealed irritation.

"Well," said Stan Ulam, "let us play a game. Imagine that we are writing a dictionary of common words. We will try to write definitions that are unmistakeably explicit, as if ready to be programmed. Let us take, for instance, nouns like 'key,' 'book,' 'passenger,' and verbs like 'waiting,' 'listening,' 'arriving.' Let us start with the word 'key.' I now take this object out of my pocket and ask you to look at it. No amount of staring at it will ever tell you that this is a key unless you already have some previous familiarity with the way keys are used.

"Now look at that man passing by in a car. How do you tell that it is not just a man you are seeing, but a passenger?

"When you write down precise definitions for these words, you discover that what you are describing is not an object, but a function, a role that is inextricably tied to some context. Take away the context, and the meaning also disappears.

"When you perceive intelligently, as you sometimes do, you always perceive a function, never an object in the set-theoretic or physical sense.

"Your Cartesian idea of a device in the brain that does the registering is based upon a misleading analogy between vision and photography. Cameras always register objects, but human perception is always the perception of functional roles. The two processes could not be more different.

"Your friends in A.I. are now beginning to trumpet the role of contexts, but they are not practicing their lesson. They still want to build machines that see by imitating cameras, perhaps with some feedback thrown in. Such an approach is bound to fail since it starts out with a logical misunderstanding."

By this time I had no choice but to agree. However, out of curiosity I decided to play devil's advocate and watch his reaction.

"But if what you say is right, what becomes of objectivity, an idea that is so definitively formalized by mathematical logic and by the theory of sets on which you yourself have worked for many years in your youth?"

There was visible emotion in his answer. "Really? What makes you so sure that mathematical logic corresponds to the way we think? You are suffering from what the French call a *déformation professionnelle*. Look at that bridge over there. It was built following logical principles. Suppose that a contradiction were to be found in set theory. Do you honestly believe that the bridge might then fall down?"

"Do you then propose that we give up mathematical logic?" said I, in fake amazement.

"Quite the opposite. Logic formalizes only very few of the processes by which we think. The time has come to enrich formal logic by adding some other fundamental notions to it. What is it that you see when you see? You see an object *as* a key, you see a man in a car *as* a passenger, you see some sheets of paper *as* a book. It is the word 'as' that must be mathematically formalized, on a par with the connectives 'and,' 'or,' 'implies,' and 'not' that have already been accepted into a formal logic. Until you do that, you will not get very far with your A.I. problem."

"This sounds like an impossible task," said I, trying to sound alarmed.

"It is not so bad. Recall another instance of a similar situation. Around the turn of the century, most mathematicians and physicists probably thought that the common sense notion of simultaneity of events was an obvious one, and needed no exact explanation. Yet a few years later Einstein came out with his special theory of relativity which exactly analyzed it. I see no reason why we shouldn't do with 'as' what Einstein did with simultaneity.

"Do not lose your faith," concluded Stan. "A mighty fortress is our mathematics. Mathematics will rise to the challenge, as it always has."

At this point, Stan Ulam started walking away, and it struck me that he was doing precisely what Descartes and Kant and Charles Saunders Peirce and Husserl and Wittgenstein had done before him at a similar juncture. He was changing the subject.

We never returned to this topic in our conversations, and since then, whenever I recall that day, I wonder whether or when A.I. will ever crash the barrier of meaning.

CHAPTER V

Stan Ulam

May 9, 1984

Stan Ulam resented being labeled an intellectual. He would not even agree to being classified as a mathematician. He referred to the published volume of his scientific papers as "a slim collection of poems." Throughout his life, his style in speaking and writing remained the aphorism, the lapidary definition, the capture of a law of nature between one subject and one predicate. "Whatever is worth saying can be stated in fifty words or less," he used to admonish us, and to teach us by his example.

Mathematics is a cruel profession. Solving a mathematical problem is for most mathematicians an arduous and lengthy process which may take years, even a lifetime. The final conquest of the truth comes, if ever, inevitably tinged with disillusion, soured by the realization of the ultimate irrelevance of all intellectual endeavor. For Stan Ulam, this process took place instantaneously, in a flash, unremittingly, day and night, as a condition of his being in the world. He was condemned to seeing the truth of whatever he saw. His word then became the warning of the prophet, the mumbled riddle of the Sybil in her trance. The comforts of illusion were denied to him.

His eyesight followed the bidding of his mind. He could focus on a detail so small as to have been missed by everyone. He could decipher a distant rumbling that no one could yet be aware of. But his blindness for average distances kept him from ever enjoying some rest in the quiet lull of mediocrity.

Worried that we might not be ready to bear the burden of those visions, he solicitously improvised daily entertainment, games into which he prodded us all to be players, flimsy amusements and puzzles he built out of his concern that we be spared, as he could not be, the sight of the naked truth. What saved him, and what was to immensely benefit science, was his instinct for taking the right step at the right time, a step which he invariably took with a scintillating display of elegance.

The inexorable laws of elegant reasoning, which he faithfully observed, became his allies as he drew out the essentials of a new idea, a gem of the mind that he would casually toss off at the world, always at the right time, when ready to be pursued and developed by others. His ideas have blossomed into theories that now grace the world of science. Measurable cardinals have conquered set theory; his foundations of probability have become bedrock. He invented more than one new stochastic process, starting from the imaginary evidence he alone saw beyond the clumsy array of figures spewed out by the very first computers. The strange recurrences of the dynamical systems he was first to describe and simulate are the key to the new dynamics of today.

Stan Ulam came to physics comparatively late in life. With unerring accuracy, he zeroed in on the one indispensable item in the baggage of the physicist, the ability to spot and shake the one essential parameter out of a morass of data. In his work at the Los Alamos Laboratory, he was the virtuoso who outguessed nature, who could compute physical constants old and new to several decimal places, guided only by an uncanny sense for relative orders of magnitude.

Every day at dawn, when most of New Mexico was asleep, Stan Ulam would sit in his study in Santa Fe and write out cryptic outlines on small pieces of paper, often no larger than postage stamps. Rewritten, reformulated, rebroadcast by others to the four corners of the earth, these notes became the problems in mathematics that set the style for a whole period. To generations of mathematicians, Ulam's problems were the door that led them into the new, to the first sweet taste of discovery.

I wish we could have convinced him that his problems will last longer than he expected them to, that they are and will be the source of much mathematics that is and will be made, that he will still find them around in a next life, sprinkled in the research papers and in the textbooks of whatever time; to convince him that they will brighten our lives, and the lives of those who come after us, like a cascade of stars in the crystal sky of Los Alamos, like the fireworks of the Fourth of July.

CHAPTER VI

The Lost Café

On doit des égards aux vivants; on ne doit aux morts que la vérité.

Voltaire

One morning in 1946 in Los Angeles, Stanislaw Ulam, a newly appointed professor at the University of Southern California, awoke to find himself unable to speak. A few hours later he underwent dangerous surgery after the diagnosis of encephalitis. His skull was sawed open and his brain tissue was sprayed with antibiotics. After a short convalescence he managed to recover apparently unscathed.

In time, however, some changes in his personality became obvious to those who knew him. Paul Stein, one of his collaborators at the Los Alamos Laboratory (where Stan Ulam worked most of his life), remarked that while Stan had been a meticulous dresser before his operation, a dandy of sorts, afterwards he became visibly sloppy in the details of his attire even though he would still carefully and expensively select every item of clothing he wore.

Soon after I met him in 1963, several years after the event, I could not help noticing that his trains of thought were not those of a normal person, even a mathematician. In his conversation he was livelier and wittier than anyone I had ever met; and his ideas, which he spouted out at odd intervals, were fascinating beyond anything I have witnessed before or since. However, he seemed to studiously avoid going into any details. He would dwell on any subject no longer than a few minutes, then impatiently move on to something entirely unrelated.

Out of curiosity, I asked John Oxtoby, Stan's collaborator in the thirties (and, like Stan, a former Junior Fellow at Harvard) about their working habits before his operation. Surprisingly, Oxtoby described how at Harvard they would sit for hours on end, day after day, in front of the blackboard. From the time I met him, Stan never did anything of the sort. He would perform a calculation (even the simplest) only when he had absolutely no other way out. I remember watching him at the blackboard, trying to solve a quadratic equation. He furrowed his brow in rapt absorption while scribbling formulas in his tiny handwriting. When he finally got the answer, he turned around and said with relief: "I feel I have done my work for the day."

The Germans have aptly called *Sitzfleisch* the ability to spend endless hours at a desk doing gruesome work. *Sitzfleisch* is considered by mathematicians to be a better gauge of success than any of the attractive definitions of talent with which psychologists regale us from time to time. Stan Ulam, however, was able to get by without any *Sitzfleisch* whatsoever. After his bout with encephalitis, he came to lean on his unimpaired imagination for his ideas, and on the *Sitzfleisch* of others for technical support. The beauty of his insights and the promise of his proposals kept him amply supplied with young collaborators, willing to lend (and risking the waste of) their time.

A crippling technical weakness coupled with an extraordinary creative imagination is the drama of Stan Ulam. Soon after I met him, I was made to understand that, as far as our conversations went, his drama would be one of the Forbidden Topics. Perhaps he discussed it with his daughter Claire, the only person with whom he would have brutally frank verbal exchanges; certainly not with anyone else. But he knew I knew, and I knew he knew I knew.

Stanislaw Ulam was born about 1907 (a date that varied as the years went by) into a family that stood as high on the social ladder as a Jewish family could at the time. He was the golden boy from a wealthy family of Lwow, Poland. In that region of Central Europe, the Ulam name was then synonymous with banking wealth: his uncles owned one of the more prosperous banks. He was educated by private tutors and in the best schools. As a child, he already showed an unusual interest in astronomy ("I am starstruck," he would often tell me as we were walking in the starry nights of Santa Fe) and in physics. In high school he was a top student, far too bright for his age. His quick wit got him good grades with little effort, but gave free rein to the laziness that was to overwhelm him in his later years.

The two authors he thoroughly read in his teens were Karl May and Anatole France. They had a formative influence on his personality, and throughout his life he kept going back to them for comfort. From Karl May's numerous adventure novels (popular enough in the German-speaking world to be among Hitler's favorite books) he derived a childlike and ever-fresh feeling of wonder, as is often found in great men. From Anatole France he took his man-of-the-world mannerisms, which in later life would endear him to every young person he spoke to.

He kept a complete set of Karl May's novels (in German, the language of his childhood) on a shelf behind his desk until he died. He regretted that a Pléiade edition of Anatole France had not been published, which he could keep by his bedside (ironically, a Pléiade volume of Anatole France's novels was published shortly after Ulam's death). He often gave me paperbacks of Anatole France's novels, which he had bought for me on one of his frequent trips to Paris. He would dedicate them to me with amusing inscriptions, urging me to read them. I regret to admit I still haven't read them.

There was never any doubt that he would study mathematics when at age seventeen he enrolled at Lwow Polytechnic Institute. Shortly after classes started, he discovered with relief that the mathematics that really mattered was not taught in the classroom but was to be found alive in one of the cafés in town, the Scottish Café. There the Lwow mathematicians would congregate daily. Between a shot of brandy and a cup of coffee, they would pose (and often solve) what turned out to be some of the outstanding conjectures in mathematics of their time, conjectures that would be dashed off on the marble of coffee tables in late evenings, in loud and uninhibited brawls.

The Lwow school was made up of offbeat, undisciplined types. Stan's teacher Stefan Banach was an alcoholic, and his best friend Mazur was a communist agitator. They cultivated the new fields of measure theory, set theory and functional analysis, which required little background and a great deal of chutzpah. The rival Warsaw school, more conservative, looked down on the Lwow mathematicians as amateurish upstarts, but the results of the Lwow school soon came to be better known and appreciated the world over, largely after the publication of Banach's book *Théorie des Opérations Linéaires*,[1] in which Ulam's name is frequently mentioned.

One day the amateur Ulam went one up on the Warsaw mathematicians, who were cultivating the equally new field of algebraic topology. While chatting at the Scottish Café with Karol Borsuk (an outstanding Warsaw topologist), he saw in a flash the truth of the Borsuk-Ulam theorem (as it is now called). Borsuk had to commandeer all his technical resources to prove what Stan Ulam had guessed. News of the result quickly swept across the ocean, and Ulam became an instant topologist.

Stan took to café mathematics like a duck to water. He quickly became the most daring of the Lwow mathematicians in formulating bold new mathematical conjectures. Almost all his guesses have been proved true, and are now to be found as theorems scattered in graduate textbooks.

In the casual ambiance of the Scottish Café, Stan blossomed into one of the most promising mathematicians of his generation. He also began to display the contradictory traits in his behavior that after his operation were to become dominant: deep intuition and impatience with detail, playful inventiveness and dislike of prolonged work. He began to view mathematics as a game, one that a well-bred gentleman should not take seriously. His insights have opened new areas of mathematics, still actively cultivated. But he himself could not bear to look upon his discoveries with more than a passing interest, and in his bitter moments he would make merciless fun of those who did take them too seriously.

The only papers in mathematics he wrote by himself date back to his early days in Lwow. They were mostly written in one sitting, often in the long stretch of a night's work, probably in response to some colleague's challenge at the Scottish Café. Much of his present reputation as a mathematician rests on these early short brilliant papers published in the Polish journal *Fundamenta Mathematicae*, which would become renowned wherever mathematics is done. His measurable cardinals, the best idea he had in this period, are still the mainspring of much present work in set theory. More often however, his flashes of originality, scattered as they are in odd contexts, have been appropriated by others with little acknowledgement, and they have proved decisive in making more than one career in mathematics. For example, his paper with Lomnicki on the foundations of probability, which also dates back to his Polish period, contains a casual remark about the existence of prime ideals in Boolean algebras, later developed by Alfred Tarski and others in several formidable papers.

The Borsuk-Ulam theorem was striking enough to catch the attention of Solomon Lefschetz. Through Lefschetz' influence, Ulam was invited in 1936 to visit the Institute for Advanced Study in Princeton. He was assigned to be John von Neumann's assistant.

For three years he commuted between Poland and America, first to Princeton and then to the Society of Fellows at Harvard, while living in luxury on his parents' monthly checks. In the summer of 1939, shortly after he returned to the U.S. in the company of his brother from what was to be his last visit to his family, World War II broke out. By accident, he had been saved from almost certain extinction. He would never leave the U.S. again, except on short trips.

The *belle époque*, the period that runs between 1870 and World War I (though some claim it ended later), was one of the happiest times of our civilization. Vienna, Prague, Lwow, Budapest were capitals of turn-of-the-century sophistication, though they lacked the staid tradition of Paris, Florence, or Aranjuez. Musil, Mahler, Kafka, Wittgenstein and the philosophers of the Vienna Circle have become for us symbols of "*mitteleuropäische Kultur.*" Most of these now legendary figures betrayed personality traits similar to Ulam's: restlessness, intolerance, a dialectic of arrogance and contrition, and an unsatisfied need for affection, compounded by their society's failure to settle on a firm code for the expression of emotion. Perhaps the tragedy that befell Central Europe could be traced back to these men's tragic lives and their repressed personalities rather than to the scurrilous outbursts of some demented house painter. When the catastrophe came, those who were alive and watched their world go up in flames remained emotionally crippled for the rest of their lives, never recovering from the shock.

Stan Ulam was one of them. Had he been able to remain in Poland and survive the War (as Steinhaus, Kuratowski, and other Jews did in hiding), he could have gone on to become one of the leading international figures of pure mathematics, at least on a par with Banach. But after he bid farewell to his friends at the Scottish Café, something died forever within him, and his career as a pure mathematician went permanently adrift.

Like other spoiled immigrants from the European bourgeoisie, Stan arrived in the United States ill-equipped for the rigors of American puritanism. The wide open spaces of America, the demands for aloneness and self-reliance that this country makes on everyone made him feel estranged. He wished to belong in the U.S., and he loved this country, but he never came to feel fully at home in it, whether in Cambridge, Madison, or Los Alamos. He missed the street life of European cities, the rambling conversations (what the Spanish call *tertulias*); he viewed with alarm the decay of an art which in our day has become all but extinct. Nowadays, the effective American way of scientific exchange has imposed itself on the rest of the world, in the seminar and in the classroom. But fifty years ago life in American universities was incomparably duller than the café-science of Lwow. Cambridge in the thirties was too cold, and, what was worse, there were no cafés. And then the War started.

In the fall of 1939, Stan spent endless hours watching the Charles River from his room in one of the newly inaugurated Harvard colleges (Eliot House I believe), stupefied by the sudden turn of events that had changed his life and those of so many others. He learned of the fall of Poland, of his family's being deported to a concentration camp (both his father and his sister were killed in gas chambers), and of the great Ulam bank being ransacked by the Soviets in one day.

He was alone now. His father's monthly checks stopped coming. His Junior Fellowship would soon run out, and he would have to support his brother's college education at Brown. He pinned great hopes on his big paper in ergodic theory,[2] which he had just finished writing with Oxtoby, and which had been accepted for publication in the *Annals of Mathematics*, the most prestigious mathematics journal of the time. Despite his personal problems, he was still delivering brilliant lectures at Harvard on the theory of functions of several real variables (my freshman advisor at Princeton, Roger Lyndon, once told me that he cherished the notes he had taken in that course).

G. D. Birkhoff, the ranking Harvard mathematician and the absolute monarch of American mathematics, took a strong liking to Stan Ulam. Like other persons rumored to be anti-Semitic, G. D. Birkhoff would feel the urge to shower his protective instincts on some good-looking young Jew. Stan Ulam's sparkling manners were diametrically opposite to Birkhoff's hard-working, aggressive, touchy personality. Birkhoff tried to keep Ulam at Harvard, but his colleagues balked at the idea. After all, Ulam had only one long paper in course of publication, and it can be surmised that the Harvard mathematicians of the thirties turned their noses at the abstract lucubrations of a student of Banach.

Birkhoff then began to write letters to his friends at several universities, suggesting Ulam's name for appointment. It didn't take long before Stan received a handsome offer from the University of Wisconsin in Madison, an assistant professorship carrying an unusually high stipend for the time, more than three thousand dollars. He had no choice but to accept it.

For the first time in his life, Stan now had to do an honest day's work, and he didn't like the idea. The teaching load of some twelve hours a week of pre-calculus soon turned into a torture. Rumor had it that he had once fallen asleep while lecturing. Madison, Wisconsin, a friendly Midwestern town, was the end of the world for a jaded young European. The ambiance was more non-existent than dismal. His colleagues, upright men (and later world-renowned mathematicians) like C. J. Everett and Steve Kleene, were not the garrulous Slavic types who might turn him on. Luckily for Stan, after his second year in Wisconsin America entered the War.

This time, it was John von Neumann who came to Stan's rescue.

Of all escapes from reality, mathematics is the most successful ever. It is a fantasy that becomes all the more addictive because it works back to improve the same reality we are trying to evade. All other escapes — sex, drugs, hobbies, whatever — are ephemeral by comparison. The mathematician's feeling of triumph, as he forces the world to obey the laws his imagination has freely created, feeds on its own success. The world is permanently changed by the workings of his mind and the certainty that his creations will endure renews his confidence as no other pursuit. The mathematician becomes totally committed, a monster, like Nabokov's chess player who eventually sees all life as subordinate to the game of chess.

Many of us remember the feeling of ecstasy we experienced when we first read von Neumann's series of papers on rings of operators in Hilbert space. It is a paradise from which no one will ever dislodge us (as Hilbert said of Cantor's set theory). But von Neumann's achievements went far beyond the reaches of pure mathematics. Together with Ulam, he was the first to have a vision of the boundless possibilities of computing, and he had the resolve to gather the considerable intellectual and engineering resources that led to the construction of the first large computer, the Maniac. No other mathematician in this century has had as deep and lasting an influence on the course of civilization.

Von Neumann was a lonely man with serious personal problems. His first wife ran away with a graduate student. He had trouble relating to others except on a strictly formal level. Whoever spoke to him noticed a certain aloofness, a distance that would never be bridged. He was always formally dressed in impeccable business suits, and he always kept his jacket on (even on horseback), as if to shield himself from the world.

Soon after Stan was appointed his assistant at the Institute for Advanced Study in Princeton in 1936, they became close friends. A similar background and a common culture shock brought them together (von Neumann came from a wealthy family in Budapest, ennobled by the Hapsburg dynasty). They would spend hours on end in silly giggles and gossip, swapping Jewish jokes, and drifting back to mathematical talk.

Stan was the only close friend von Neumann had. Von Neumann latched on to Stan and managed to be close to him as often as possible. Stan was the more original mathematician of the two, though he accomplished far less in mathematics than von Neumann did; Von Neumann had an incomparably stronger technique. From their free play of ideas came some of the great advances in applied mathematics: the Monte Carlo method, mathematical experiments on the computer, cellular automata, simulated growth patterns.

Stan was somewhat embarrassed at first by von Neumann's displays of friendship (von Neumann was six years older). Following his uncanny instinct for doing the right thing at the right time, Stan soon found the way to cheer up his brooding friend. He began to make fun of von Neumann's accomplishments. He would mercilessly ridicule continuous geometries, Hilbert space and rings of operators. Like everyone who works with abstractions, von Neumann needed constant reassurance against deep-seated and recurring self-doubts. Stan Ulam's jibes would cleverly pick on weaknesses in von Neumann's work that were obvious and expected. They were an indirect but firm expression of admiration. Rather than feeling offended, von Neumann would burst out in a laughter of relief.

Much later, when Stan related these events to me, he affected to regret never having said a kind word to von Neumann about his work in pure mathematics. But he was not serious. Deep inside, he knew he had been good to his friend.

Stan didn't fully realize how much von Neumann meant to him until von Neumann began to die of cancer in 1955. Stan would visit him every day at Walter Reed Hospital in Washington, ready with a bagful of the latest jokes and prurient Los Alamos gossip. The little hospital bed would shake with the vibrations of von Neumann's big belly as he laughed himself to tears, the very tears that Stan was fighting to control. After von Neumann died, Stan cried in the corridor outside the room, oblivious to anyone present. It was probably the only time in his life when he publicly lost control of his emotions.

Shortly after the U.S. entered the War in 1941, Stan (at Wisconsin) began to notice that von Neumann's letters were becoming infrequent. Curious about his friend's mysterious unavailability, Stan managed one day to corner him in Chicago, and then he got wind that a big project was in the offing, in which the best scientific minds in the country would be involved. He implored von Neumann to drag him out of his Wisconsin rut and to get him a job with the Manhattan Project (as it came to be called shortly afterwards). Von Neumann had probably already made up his mind to bring Stan with him to the newly founded Los Alamos Laboratory, where the atomic bomb project was being launched, and was glad to comply with the request.

The choice of a set theorist for work in applied physics might seem eccentric, but in retrospect, von Neumann turned out to have made the right decision. Besides, he felt even lonelier in Los Alamos than he did in Princeton, as the token mathematician in a sea of physicists (though he was probably one of the two or three finest minds among them, together with Enrico Fermi and Richard Feynman).

The assembly of geniuses who roamed the corridors of the Los Alamos Lab during World War II has not been matched in recorded history, with the possible exception of ancient Greece. In the hothouse of the Manhattan Project, Stan's mind opened up again, as it hadn't since the days of the Scottish Café. The joint efforts of the best scientists, their talents stimulated and strained by the challenge of a difficult project, made what could have been a drab weapons laboratory into a cradle of new ideas. In conspiratorial whispers between long stints at the bench, in a corner at some loud drinking party, the postwar revolutions in science were being hatched.

Los Alamos was a turning point in Stan Ulam's career. Physics, not mathematics, became the center of his interest. After carefully watching Fermi and Feynman at the blackboard, he discovered that he too had a knack for accurately estimating physical quantities by doing simple calculations with orders of magnitude. In fact, he turned out to be better at that game than just about anyone around him.

It is hard to overstate how rare such an ability is in a mathematician. The literalness of mathematics is as far removed from the practical needs of the physicist as might be the story of the Wizard of Oz. As Stan giddily began to display his newly found talent, he came to rely less and less on standard mathematical techniques, and to view ordinary mathematics with downright contempt. He admired Fermi's genius for solving physical problems with no more than the minimum amount of math. From that time, Fermi remained for him the ideal of a scientist. In his old age, he liked to repeat with bitterness (and perhaps with a touch of exaggeration) that Fermi had been the last physicist.

But the Magic Mountain lasted only as long as the War. In 1945 it looked as if the Los Alamos Lab might close down, like many other war projects, and Stan began to look for a job elsewhere. His list of publications was no longer than it had been in 1939, and unpublished work gets no credit. To his chagrin, he was turned down by several of the major universities, beginning with Harvard. Von Neumann's strong backing did not help (perhaps because everyone knew they were friends). He had to accept the offer of a professorship at the University of Southern California, a second-rate institution but one with great plans for the future.

Suddenly he found himself in the middle of an asphalt jungle, teaching calculus to morons. The memories of his friends in Los Alamos, of the endless discussions, of the all-night poker games, haunted him as he commuted daily among the wilting palm trees of Los Angeles. The golden boy had lost his company of great minds, his audience of admirers. Like anguish that could no longer be contained, encephalitis struck.

We tend to regard disease as a physical occurrence, as an unforeseen impairment of the body that also, mysteriously, affects the mind. But this is an oversimplification. After a man's death, at the time of the final reckoning, an event that might once have appeared accidental is viewed as inevitable. Stan Ulam's attack of encephalitis was the culmination of his despair.

After recovering from his operation he resigned his position in a hurry and went back to Los Alamos. The year was 1946, and the Los Alamos Lab had become a different place. Gone were most of the great minds (though many of them would make cameo appearances as consultants); and since the success of the atomic bomb, the federal government was lavishing limitless funds on the lab (Norris Bradbury, the director, would return unused funds to the Treasury at the end of each fiscal year). For a few years Los Alamos scientists found themselves coddled, secure and able to do or not do whatever they pleased, free to roam around the world on red-carpeted military flights (that is, until Americans decided to give up the empire they had won).

Ulam came back to Los Alamos haunted by the fear that his illness might have irreparably damaged his brain. He knew his way of thinking had never been that of an ordinary mathematician, and now less than ever. He also feared that whatever was left of his talents might quickly fade. He decided the time had come to engage in some substantial project that would be a fair test of his abilities, and to which his name might perhaps remain associated.

While at Wisconsin, he had met C. J. Everett. They had jointly written the first paper in the subject that is now called algebraic logic[3] (a beautiful paper that has been plundered without acknowledgement). Everett, a reclusive and taciturn man, was richly endowed with the ability to compute. He was a good listener, and after he was appointed a staff member at the Los Alamos Laboratory he constantly suffered from a paranoid fear of being fired for wasting lab time on research in pure mathematics. This fear eventually led him to an early retirement. Everett turned out to be a perfect complement to Stan. After he had accepted Stan's invitation to join the Los Alamos Lab, he joined forces with Stan on a long and successful collaboration.

As their first project, they chose the theory of branching processes. They believed they were the first to discover the probabilistic interpretation of functional composition. They ignored all previous work, all the way back to Galton and Watson in the nineteenth century. Stan never had the patience to leaf through published research papers. He hated to learn from others what he thought he could invent by himself — and often did.

They rediscovered all that had been already done, and added at least as much of their own. Their results were drafted by Everett in three lengthy lab reports, which found substantial applications in the theory of neutron diffusion, an essential step in the understanding of nuclear reactions. These reports were never published during the authors' lifetimes, but they nevertheless had a decisive influence on the development of what continues to be a thriving area of probability theory. The authors have received little acknowledgement for their work, perhaps as a spiteful punishment for their own neglect of the work of others.

Their second project was the hydrogen bomb.

Stan Ulam and Edward Teller disliked each other from the moment they met. Since the days of the Manhattan Project, Teller had been something of a loner. He was never quite accepted by the mainline Bethe-Fermi-Oppenheimer group as one of them (as Stan was), and not even his fellow Hungarian von Neumann felt at ease with him. This, despite the fact that Teller distinguished himself from the first days of Los Alamos as one of the most brilliant applied physicists (one of the best of the century).

Teller related with difficulty and diffidence to other scientists of his age. He felt more at ease either with young people, or with celebrities, highly placed politicians, generals and admirals. His group (what became the Lawrence Livermore National Laboratory after he left Los Alamos in a huff) was highly disciplined, rank-conscious and loyal. He would sagely guide his students and assistants to doing the best research work they were capable of, and he would reward his creatures with top-rank positions in academic administration or in government, all the way up to Secretary of Defense, President of Caltech, and Director of the Salk Institute.

Since the success of the atomic bomb, Teller had been obsessed by the idea of the "Super" (later called the hydrogen bomb). Because of disagreements between him and Oppenheimer (the first director of the Los Alamos Laboratory), his project had more than once been on the verge of being canceled. Now, after the brouhaha about Oppenheimer's clearance that was to cast Oppenheimer as a martyr and Teller in the role of bad guy, Stan Ulam was out to get Teller by proving that Teller's plans for the new bomb could not work.

Everett and Ulam worked frantically in competition with Teller's group. They met in the morning for several hours in a little office out of the way. Ulam would generate an endless stream of ideas and guesses and Everett would check each one of them with feverish computations. In a few months, Everett wore out several slide rules. Stan's imagination roamed wild and gleeful as he devised a thousand tricks to make the first hydrogen weapon work. One of them worked.

Of all mental faculties, shrewdness is the most stable, and the last to go. To the end of his days, Stan Ulam was able to size up with unerring accuracy situations and turns of events that might accrue in his favor. He now decided that his work on the hydrogen bomb was such an opportunity, the best ever to come along.

The full extent of his contribution to the design of the first hydrogen bomb may never be precisely established. It is certain, however, that he was instrumental in demolishing several mistaken proposals that might otherwise have resulted in considerable waste of time and funds. The loud dispute with Teller over the priority of the idea that made the hydrogen bomb work, which Stan slyly encouraged, brought him wide publicity (the patent application for the device is jointly signed by Teller and Ulam). Teller was at the time picking up allies among conservative groups. The Democrats saw an advantage to adopting his rival Ulam as their champion of liberalism in science. He was invited to sit in on important Washington committees. He became a darling of the Kennedy era, and a representative figure of science.

At last, Stan found some of the lost glitter of his Polish youth, if not in the form of tangible wealth, at least in the guise of worldwide recognition. I have never had a discussion with Stan of the wider issues connected with the making of atomic devices, nor did he encourage such discussion. Although he claimed to have witnessed some explosions, I later discovered that he had steadfastly refused to be present at any shots, even the first one at Alamogordo, to which he had not been invited. Whatever his private feelings, he did not hesitate to claim priority for the idea, and seemed never to be bothered by any regrets.

The years after his triumph, the late forties and fifties, were the high point of Stan Ulam's life. He adjusted his personality to his new role as an international celebrity and as the resident genius of the Los Alamos Laboratory. His attention span, already short, now decreased to a dangerous low. In conversation, he had to win every argument. When he felt he was on the losing side, he would abruptly change the subject, but not before seeing the bottom of the other person's position and summarizing it with irritating accuracy. Considering how fast it all happened, it was remarkable how seldom he misunderstood.

He so liked to dominate the conversation that some ofhis colleagues would take pains to avoid him. His idea of a dialogue was to spout out a line, wait for the expected reaction of approval, and then go on to the next item on his mind, silencing with a gesture anyone who might think of butting in.

He became chillingly realistic in sizing up his peers' weaknesses. He acknowledged with resentment the fact that Feynman's and Ted Taylor's were superior minds. But by and large, he would pass judgment on others by what they had done, and on himself by what he thought he could yet accomplish.

The free rein Ulam gave to his fantasy fed on one of his latent weaknesses: his wishful thinking. He became an artist of self-deception. He would go to great lengths to avoid facing the unpleasant realities of daily life. When anyone close to him became ill, he would clutch at every straw to pretend that nothing was really wrong. When absolutely forced to face an unpleasant event, he would drop into a chair and fall into a silent and wide-eyed panic.

Despite the comfort of the Los Alamos Lab (in the fifties and sixties Ulam was one of two research advisors to the Director of the lab), Stan could not find peace there. He became a permanent traveler. Ever since he returned to Los Alamos in 1946 he had lived, unbelievable as it may sound, out of a suitcase. He thought of himself as permanently on the road. Why settle down anywhere? The Scottish Café was gone forever anyway. He was a passenger on an imaginary ship who survived on momentary thrills designed to get him through the day. He surrounded himself with traveling companions who were fun to be with and talk to. He went to any lengths to avoid being alone. When absolutely forced to be by himself (early in the morning at home, or in some hotel room) he would concentrate on mathematics as he seldom did, to avoid heeding the clamor of his memories.

I will always treasure the image of Stan Ulam sitting alone in his study in Santa Fe early in the morning, rapt in thought, scribbling formulas in drafts that would fill a couple of postage stamps.

The traits of Stan Ulam's personality that became dominant in his later years were laziness, generosity, considerateness, and most of all, incisiveness of thought.

Those who knew Stan and did not know what to make of him covered up the mixture of envy and resentment they felt toward him by pronouncing him lazy. He was proverbially lazy. In the thirties, in Cambridge, he would take a taxi to Harvard whenever he spent a night in Boston, to avoid tackling the petty decisions that a ride on the subway required. He was fond of telling me that once, as his taxi crossed the Longfellow Bridge, he caught a glimpse of President Lowell going to work, hanging on a strap in a crowded car, and he blushed.

One day in the sixties I found him lying on the sofa in his living room in Santa Fe with the day's newspaper underneath him. He tore off a piece of paper at a time, read it and threw it to the floor. He would not bring himself to make the effort of standing up and removing the paper from his back.

There is a spot on a pathway up the mountains from Los Alamos that is called "Ulam's Landing." It is as far as Stan ever went on a hike before turning back. More often, he would watch the hikers with binoculars from the porch of his house while sipping gin and tonics and chatting with friends.

Like all words denoting human conditions, laziness, taken by itself, is neutral. Each person is forced to invent whatever sense his or her own laziness will have in his life. Stan Ulam turned his laziness into a challenge. He had to give his thinking an epigrammatic twist of elegant definitiveness. His laziness became an imperious demand to get to the heart of things with a minimum of jargon. On any subject, however little his competence, he could contribute some valuable original remark. After tossing a brilliant gem at whoever might be listening and satisfying himself that his latest fantasy made some sense, he would impatiently move on to whatever came next to mind, leaving to his audience all further work on the previous subject.

He had a number of abrupt conversation stoppers to get rid of bores. One of them was a question which he used to stop some long tirade: "What is this compared to $E = mc^2$?" When I first heard it (undoubtedly it was being used to stop me) I thought it a sign of conceit. But I was wrong. He would wake up in the middle of the night and compare his own work to $E = mc^2$, and he developed ulcers from these worries. His apparent conceit was a way of concealing from others, and most of all from himself, an impairment of his brain that was constantly growing worse. On rare occasions he felt overwhelmed by guilt at his inability to concentrate, which he viewed as avoidance of "serious" work. He looked at me, his glaucous eyes popping and twitching (they were the eyes of a medium, like Madame Blavatzky's), his mask about to come down, and asked, "Isn't it true that I am a charlatan?" I proceeded to set his mind at rest by giving him, as a sedative, varied examples of flaming charlatans taken from scientists we both knew (with and without Nobel prizes). But soon his gnawing doubts would start all over. He knew he would remain to the end a Yehudi Menuhin who never practiced.

His generosity was curiously linked to his laziness. A generous action is often impulsive, and calls for little foresight. Its opposite requires the careful advance planning that Stan loathed. He fancied himself a *grand seigneur* of bottomless means, and in matters of money he regularly practiced the art of selfdeception. He would go to great lengths to conceal his shaky financial condition. He lived beyond his means. He would invite a half-dozen friends to dinner on the spur of the moment, sometimes without warning his wife Françoise. He carried on his person bundles of fifty and one hundred dollar bills, partly out of a remnant of refugee mentality, partly to impress whomever he met on his travels by paying for lavish dinners and passing out presents to his young friends.

He was also too proud to insist on his priority for the new ideas he contributed to science. His nonchalance as to the fate and success of his work has unjustly lowered his standing as a scientist. When he saw one of his ideas circulating without credit, he remarked, "Why should they remember me? No one quotes Newton or Einstein in the bibliography of their papers."

His way of expressing himself lent itself to exploitation. He would spout out incoherent pronouncements that seemed to make little sense, deliberately couched in hieratic duplicity. Those of his listeners who decided to pursue his proposals (and who often ended up writing research papers on them) felt they had made enough of an effort figuring out what Stan really meant, and rewarded themselves by claiming full credit.

A seed idea is the last thing we acknowledge, all the more so when it originates from a native intelligence blessed with inexhaustible luck. After we silently appropriate it, we figure out soon enough a way to obliterate all memory of its source. In a last-ditch effort to salvage our pride, we will manage to find fault with the person to whom we are indebted. Stan Ulam's weaknesses were all too apparent, and made him more vulnerable than most. But the strength of his thinking more than made up for what he lost to the pettiness of his underlings.

Stan did his best work in fields where no one dared to tread, where he would be sure of having the first shot, free from fear of having been anticipated. He used to brag about being lucky. But the source of his luck was his boundless intellectual courage, which let him see an interesting possibility where everyone else had only seen a blur.

He refused to write down some of his best ideas. He thought he would someday find the time and the help he needed to work them out. But he was misjudging the time he had left. His best problems will survive only if his students ever write them down.

Two of them have struck me. In the nineteeth century, mathematicians could not conceive of a surface unless defined by specific equations. After a period of abstraction, the point-set topologists in this century arrived at the abstract notion of a topological space, which renders in precise terms our intuitive grasp of the notion of extension. Ulam proposed to go through a similar process of refinement starting with Maxwell's equations, to arrive at an appropriate abstract structure for electromagnetic theory that would be free of algebraic irrelevancies.

The second bore on ergodic dynamical systems. Poincaré and several others after him taught us that in such a system every state is visited infinitely often, given a sufficiently long time. In practice, however, the recurrence times are so long that one cannot observe successive visits, and the practical import of ergodicity is nil. This paradox became strikingly evident after the Fermi-Pasta-Ulam simulation experiments of coupled harmonic oscillators (written up in one of Fermi's last papers; Stan told me that Fermi considered this to have been his most important discovery). In these non-linear systems, the initial state is visited several times before another set of available states is even approached.

After observing this phenomenon, Ulam guessed that in some ergodic systems the phase space ought to be measure-theoretically represented by two or more big blobs connected by thin tubes. He wanted to express his guess in terms of ergodic theory. I wish we knew how.

In his fascination with physics, he was led to formulate mathematical thoughts that had a background of physics, but that bore the unmistakable ring of mathematics (he once started to draft a long paper which was to be titled "Physics for Mathematicians"). One of the most striking is his proposal for the reconstruction of the CGS system on the basis of random walk. Another is the existence of a limit distribution in the redistribution of energy by collision of particles, which R. D. Mauldin proved true.

Stan Ulam's best work in mathematics is a game played in the farthest reaches of abstraction, where the cares of the world cannot intrude: in set theory, in measure theory, and in foundations. He referred to his volume of collected papers as a slim volume of poems. It is just that.

He would have preferred to be remembered for those of his insights that have found practical applications, such as the Monte Carlo method, where he will share the credit with Nick Metropolis and John von Neumann, and the H-bomb, where he will at best be remembered alongside Teller. Instead, his name is most likely to survive for his two problem books in mathematics, which will remain bedside books for young mathematicians, eager to make their mark by solving at least one of them.

Only in the last years of his life did his thinking take a decisively speculative turn. He professed to dislike philosophical discussions, and he excoriated ponderous treatises in philosophy. He thought them in bad taste, "Germanic" (one of his words of reprobation). Nonetheless, he had an instinctive grasp of philosophical issues, which he refused to render in words. When forced to take a philosophical stand, he would profess to agree with the naive scientism of H. G. Wells and with the positivism of the Vienna Circle (the reigning philosophy of his time), but in his thinking he was closer to Husserl's phenomenology. I gave him a book of Georg Trakl's poems; he read them and was moved to tears.

Those of us who were close to him (Al Bednarek, Bill Beyer, C. J. Everett, R. D. Mauldin, N. Metropolis, J. Mycielski, P. R. Stein and myself, to name a few) were drawn to him by a fascination that went beyond the glitter of new ideas of arresting beauty, beyond the trenchant remarks that laid bare the hidden weakness of some well-known theory, beyond the endless repertoire of amusing anecdotes. The fascination of Stan Ulam's personality rested in his supreme self-confidence. His self-confidence was not the complacency of success. It rested on the realization that the outcome of all undertakings, no matter how exalted, will be ultimate failure. From this unshakeable conviction he drew his strength.

This conviction of his was kept silent. What we heard from him instead were rambling tirades against scientists who take themselves too seriously, or against the megalomania of contemporary mathematics. He would viciously tear to shreds some of the physics that goes on today, which is nothing but poor man's mathematics, poorly learned and poorly dressed up in a phony physical language. But his faith in a few men whom he considered great remained unshaken: Einstein, Fermi, L. E. J. Brouwer, President Truman.

He could not bear to see unhappiness among his friends, and he went to any lengths to cheer us up. One day shortly before he died we were driving up the Jemez Mountains, along the stretch of straight road that starts right after the last site of the Los Alamos Lab. It is a barren road, littered with the leftovers of years of experimental explosions. I felt depressed, and drove silently, looking straight ahead.

I could feel his physical discomfort at my unhappiness. He tried telling some funny stories, but they didn't work. After a minute of silence, he deployed another tactic. He knew I had been interested in finding out how much physics he really knew, and that I had unsuccessfully tried to quiz him. Now he launched on a description of the Planck distribution (which he knew I didn't know) and its role in statistical mechanics. I turned around, surprised at the thoroughness of his knowledge, and he smiled. But a few minutes later I again fell silent, and the gloom started all over. After a pause that was undoubtedly longer than he could bear, he blurted out, "You are not the best mathematician I have ever met, because von Neumann was better. You are not the best Italian I have ever met, because Fermi was better. But you are the best psychologist I have ever met."

This time I smiled. He knew I knew he knew that I saw through his ruses better than any other person, that I admired his genius for turning his laziness into creativity, his incoherence into lapidary sentences, his brain damage into dispassionate talk. That he appreciated.

Thinking back and recalling the ideas, insights, analogies, nuances of style that I drew from my association with him for twenty-one years, I am at a loss to tell where Ulam ends and where I really begin. Perhaps this is one way he chose to survive.

One morning in May, 1984, he dropped to the floor and in one instant he was gone. Even in death he knew how to be elegant. After he died, I inherited his monocle, which perfectly fits my left eye. Too bad I won't be able to wear it: it doesn't like me. Shortly after I took it home, I locked it into a drawer in a closet; I couldn't bring myself to look at it.

After Stan died life seemed to go on as if nothing had happened, as if his daily phone calls to me were being delayed for some accidental reason (he used to make a round of phone calls to his friends every day in the late morning to relieve boredom), as if at any minute we would meet again and resume our discussions. Three years went by. One day I found the courage to unlock that drawer. The void I had felt for pretending he was still alive had become overwhelming. As I looked at the monocle, I was struck by the horror of feeling alone. From now on I would be faced with the terror of his absence.

No other period of civilization has been so dependent for survival on hypocrisy as the *belle époque*, the late post-Victorian age. It has bequeathed us a heritage of lies that we are now painfully charged with erasing, like a huge national debt. The prejudice that the scientist's search for the truth shall be free from emotion, from pettiness, from the passions of the world has been nurtured for over a century by biographers whose faith in science was too weak. One shudders to guess how many talented young minds have been deflected from science by the reading of such unrealistic portrayals of great scientists.

Fortunately, the younger generations, made prematurely street-wise (not, alas, more intelligent) by fifty years of television, will laugh at the crude obfuscations of mealy-mouthed biographers. A man's personality is no longer regarded as it was only a generation ago, as a puzzle pieced together out of components called "character traits," like a set out of elements, each piece labeled *good* or *bad*. The elements of a person's being, if any exist at all, are more likely to be ill-defined qualities like two-sidedness, ambivalence, and the tension of opposites.

Only yesterday, the drama of Stan Ulam might have been passed over in silence, whispered in the secret of the confessional, or avowed on the couch of the psychiatrist. More likely, it would have been added to a long list of forbidden subjects. In our time, it makes an edifying story. Stan Ulam was lazy, but he knew how to turn his laziness into elegance and conciseness of thought. He talked too much, but what he said was worth listening to. He was self-centered (though not egotistical), but he went farther in his thinking than anyone around him. He had an overpowering personality, but he made us into better human beings. He wasted much of his professional life, but he left us new ideas and discoveries that enrich our lives and will enrich those that come after us.

From left: Gian-Carlo Rota and Stanislaw Ulam
Santa Fe, 1974

PART II

Philosophy

A Minority View

CHAPTER VII

The Pernicious Influence of Mathematics Upon Philosophy

The Double Life of Mathematics

ARE MATHEMATICAL IDEAS INVENTED or discovered? This question has been repeatedly posed by philosophers through the ages and will probably be with us forever. We will not be concerned with the answer. What matters is that by asking the question, we acknowledge that mathematics has been leading a double life.

In the first of its lives mathematics deals with facts, like any other science. It is a fact that the altitudes of a triangle meet at a point; it is a fact that there are only seventeen kinds of symmetry in the plane; it is a fact that there are only five non-linear differential equations with fixed singularities; it is a fact that every finite group of odd order is solvable. The work of a mathematician consists of dealing with such facts in various ways. When mathematicians talk to each other, they tell the facts of mathematics. In their research, mathematicians study the facts of mathematics with a taxonomic zeal similar to a botanist studying the properties of some rare plant.

The facts of mathematics are as useful as the facts of any other science. No matter how abstruse they may first seem, sooner or later they find their way back to practical applications. The facts of group theory, for example, may appear abstract and remote, but the practical applications of group theory have been numerous, and have occurred in ways that no one could have anticipated. The facts of today's mathematics are the springboard for the science of tomorrow.

In its second life, mathematics deals with proofs. A mathematical theory begins with definitions and derives its results from clearly agreed-upon rules of inference. Every fact of mathematics must be ensconced in an axiomatic theory and formally proved if it is to be accepted as true. Axiomatic exposition is indispensable in mathematics because the facts of mathematics, unlike the facts of physics, are not amenable to experimental verification.

The axiomatic method of mathematics is one of the great achievements of our culture. However, it is only a method. Whereas the facts of mathematics once discovered will never change, the method by which these facts are verified has changed many times in the past, and it would be foolhardy to expect that changes will not occur again at some future date.

The Double Life of Philosophy

The success of mathematics in leading a double life has long been the envy of philosophy, another field which also is blessed — or maybe we should say cursed — to live in two worlds but which has not been quite as comfortable with its double life.

In the first of its lives, philosophy sets itself the task of telling us how to look at the world. Philosophy is effective at correcting and redirecting our thinking, helping us do away with glaring prejudices and unwarranted assumptions. Philosophy lays bare contradictions that we would rather avoid facing. Philosophical descriptions make us aware of phenomena that lie at the other end of the spectrum of rationality that science will not and cannot deal with.

The assertions of philosophy are less reliable than the assertions of mathematics but they run deeper into the roots of our existence. Philosophical assertions of today will be the common sense of tomorrow.

In its second life, philosophy, like mathematics, relies on a method of argumentation that seems to follow the rules of some logic. But the method of philosophical reasoning, unlike the method of mathematical reasoning, has never been clearly agreed upon by philosophers, and much philosophical discussion since its Greek beginnings has been spent on method. Philosophy's relationship with Goddess Reason is closer to a forced cohabitation than to the romantic liaison which has always existed between Goddess Reason and mathematics.

The assertions of philosophy are tentative and partial. It is not even clear what it is that philosophy deals with. It used to be said that philosophy was "purely speculative," and this used to be an expression of praise. But lately the word "speculative" has become a *bad word*.

Philosophical arguments are emotion-laden to a greater degree than mathematical arguments and written in a style more reminiscent of a shameful admission than of a dispassionate description. Behind every question of philosophy there lurks a gnarl of unacknowledged emotional cravings which act as a powerful motivation for conclusions in which reason plays at best a supporting role. To bring such hidden emotional cravings out into the open, as philosophers have felt it their duty to do, is to ask for trouble. Philosophical disclosures are frequently met with the anger that we reserve for the betrayal of our family secrets.

This confused state of affairs makes philosophical reasoning more difficult but far more rewarding. Although philosophical arguments are blended with emotion, although philosophy seldom reaches a firm conclusion, although the method of philosophy has never been clearly agreed upon, nonetheless the assertions of philosophy, tentative and partial as they are, come far closer to the truth of our existence than the proofs of mathematics.

The Loss of Autonomy

Philosophers of all times, beginning with Thales and Socrates, have suffered from recurring suspicions about the soundness of their work and have responded to them as well as they could.

The latest reaction against the criticism of philosophy began around the turn of the twentieth century and is still very much with us.

Today's philosophers (not all of them) have become great believers in mathematization. They have recast Galileo's famous sentence to read, "The great book of philosophy is written in the language of mathematics."

"Mathematics calls attention to itself,"[1] wrote Jack Schwartz in a famous paper on another kind of misunderstanding. Philosophers in this century have suffered more than ever from the dictatorship of definitiveness. The illusion of the final answer, what two thousand years of Western philosophy failed to accomplish, was thought in this century to have come at last within reach by the slavish imitation of mathematics.

Mathematizing philosophers have claimed that philosophy should be made factual and precise. They have given guidelines based upon mathematical logic to philosophical argument. Their contention is that the eternal riddles of philosophy can be definitively solved by pure reasoning, unencumbered by the weight of history. Confident in their faith in the power of pure thought, they have cut all ties to the past, claiming that the messages of past philosophers are now "obsolete."

Mathematizing philosophers will agree that traditional philosophical reasoning is radically different from mathematical reasoning. But this difference, rather than being viewed as strong evidence for the heterogeneity of philosophy and mathematics, is taken as a reason for doing away completely with non-mathematical philosophy.

In one area of philosophy the program of mathematization has succeeded. Logic is nowadays no longer part of philosophy. Under the name of mathematical logic it is now a successful and respected branch of mathematics, one that has found substantial practical applications in computer science, more than any other branch of mathematics.

But logic has become mathematical at a price. Mathematical logic has given up all claims of providing a foundation to mathematics. Very few logicians of our day believe that mathematical logic has anything to do with the way we think.

Mathematicians are therefore mystified by the spectacle of philosophers pretending to re-inject philosophical sense into the language of mathematical logic. A hygienic cleansing of every trace of philosophical reference had been the price of admission of logic into the mathematical fold. Mathematical logic is now just another branch of mathematics, like topology and probability. The philosophical aspects of mathematical logic are qualitatively no different from the philosophical aspects of topology or the theory of functions, aside from a curious terminology which, by chance, goes back to the Middle Ages.

The fake philosophical terminology of mathematical logic has misled philosophers into believing that mathematical logic deals with the truth in the philosophical sense. But this is a mistake. Mathematical logic deals not with the truth but only with the game of truth. The snobbish symbol-dropping found nowadays in philosophical papers raises eyebrows among mathematicians, like someone paying his grocery bill with Monopoly money.

Mathematics and Philosophy: Success and Failure

By all accounts mathematics is mankind's most successful intellectual undertaking. Every problem of mathematics gets solved, sooner or later. Once solved, a mathematical problem is forever finished: no later event will disprove a correct solution. As mathematics progresses, problems that were difficult become easy and can be assigned to schoolchildren. Thus Euclidean geometry is taught in the second year of high school. Similarly, the mathematics learned by my generation in graduate school is now taught at the undergraduate level, and perhaps in the not too distant future, in the high schools.

Not only is every mathematical problem solved, but eventually every mathematical problem is proved trivial. The quest for ultimate triviality is characteristic of the mathematical enterprise.

Another picture emerges when we look at the problems of philosophy. Philosophy can be described as the study of a few problems whose statements have changed little since the Greeks: the mind-body problem and the problem of reality, to mention only two. A dispassionate look at the history of philosophy discloses two contradictory features: first, these problems have in no way been solved, nor are they likely to be solved as long as philosophy survives; and second, every philosopher who has ever worked on any of these problems has proposed his own "definitive solution," which has invariably been rejected by his successors.

Such crushing historical evidence forces us to conclude that these two paradoxical features must be an inescapable concomitant of the philosophical enterprise. Failure to conclude has been an outstanding characteristic of philosophy throughout its history.

Philosophers of the past have repeatedly stressed the essential role of failure in philosophy. José Ortega y Gasset used to describe philosophy as "a constant shipwreck." However, fear of failure did not stop him or any other philosopher from doing philosophy.

The failure of philosophers to reach any kind of agreement does not make their writings any less relevant to the problems of our day. We reread with interest the mutually contradictory theories of mind that Plato, Aristotle, Kant and Comte have bequeathed to us, and find their opinions timely and enlightening, even in problems of artificial intelligence.

But the latter day mathematizers of philosophy are unable to face up to the inevitability of failure. Borrowing from the world of business, they have embraced the ideal of success. Philosophy had better be successful, or else it should be given up.

The Myth of Precision

Since mathematical concepts are precise and since mathematics has been successful, our darling philosophers infer — mistakenly— that philosophy would be better off, that is, would have a better chance of being successful, if it utilized precise concepts and unequivocal statements.

The prejudice that a concept must be precisely defined in order to be meaningful, or that an argument must be precisely stated in order to make sense, is one of the most insidious of the twentieth century. The best known expression of this prejudice appears at the end of Ludwig Wittgenstein's *Tractatus*.[2] The author's later writings, in particular *Philosophical Investigations*,[3] are a loud and repeated retraction of his earlier gaffe.

Looked at from the vantage point of ordinary experience, the ideal of precision seems preposterous. Our everyday reasoning is not precise, yet it is effective. Nature itself, from the cosmos to the gene, is approximate and inaccurate.

The concepts of philosophy are among the least precise. The mind, perception, memory, cognition are words that do not have any fixed or clear meaning. Yet they do have meaning. We misunderstand these concepts when we force them to be precise. To use an image due to Wittgenstein, philosophical concepts are like the winding streets of an old city, which we must accept as they are, and which we must familiarize ourselves with by strolling through them while admiring their historical heritage. Like a Carpathian dictator, the advocates of precision would raze the city and replace it with the straight and wide Avenue of Precision.

The ideal of precision in philosophy has its roots in a misunderstanding of the notion of rigor. It has not occurred to our mathematizing philosophers that philosophy might be endowed with its own kind of rigor, a rigor that philosophers should dispassionately describe and codify, as mathematicians did with their own kind of rigor a long time ago. Bewitched as they are by the success of mathematics, they remain enslaved by the prejudice that the only possible rigor is that of mathematics and that philosophy has no choice but to imitate it.

Misunderstanding the Axiomatic Method

The facts of mathematics are verified and presented by the axiomatic method. One must guard, however, against confusing the *presentation* of mathematics with the *content* of mathematics. An axiomatic presentation of a mathematical fact differs from the fact that is being presented as medicine differs from food. It is true that this particular medicine is necessary to keep the mathematician at a safe distance from the self-delusions of the mind. Nonetheless, understanding mathematics means being able to forget the medicine and enjoy the food. Confusing mathematics with the axiomatic method for its presentation is as preposterous as confusing the music of Johann Sebastian Bach with the techniques for counterpoint in the Baroque age.

This is not, however, the opinion held by our mathematizing philosophers. They are convinced that the axiomatic method is a basic instrument of discovery. They mistakenly believe that mathematicians use the axiomatic method in solving problems and proving theorems. To the misunderstanding of the role of the method they add the absurd pretense that this presumed method should be adopted in philosophy. Systematically confusing food with medicine, they pretend to replace the food of philosophical thought with the medicine of axiomatics.

This mistake betrays the philosophers' pessimistic view of their own field. Unable or afraid as they are of singling out, describing and analyzing the structure of philosophical reasoning, they seek help from the proven technique of another field, a field that is the object of their envy and veneration. Secretly disbelieving in the power of autonomous philosophical reasoning to discover truth, they surrender to a slavish and superficial imitation of the truth of mathematics.

The negative opinion that many philosophers hold of their own field has caused damage to philosophy. The mathematician's contempt for the philosopher's exaggerated estimation of a method of mathematical exposition feeds back onto the philosopher's inferiority complex and further decreases the philosopher's confidence.

"Define Your Terms!"

This old injunction has become a platitude in everyday discussions. What could be healthier than a clear statement right at the start of what it is that we are talking about? Doesn't mathematics begin with definitions and then develop the properties of the objects that have been defined by an admirable and infallible logic?

Salutary as this injunction may be in mathematics, it has had disastrous consequences when carried over to philosophy. Whereas mathematics *starts* with a definition, philosophy *ends* with a definition. A clear statement of what it is we are talking about is not only missing in philosophy, such a statement would be the instant end of all philosophy. If we could define our terms, then we would gladly dispense with philosophical argument.

The "define your terms" imperative is flawed in more than one way. When reading a formal mathematical argument we are given to believe that the "undefined terms," or the "basic definitions," have been whimsically chosen out of a variety of possibilities. Mathematicians take mischievous pleasure in faking the arbitrariness of definition. In fact no mathematical definition is arbitrary. The theorems of mathematics motivate the definitions as much as the definitions motivate the theorems. A good definition is "justified" by the theorems that can be proved with it, just as the proof of the theorem is "justified" by appealing to a previously given definition.

There is, thus, a hidden circularity in formal mathematical exposition. The theorems are proved starting with definitions; but the definitions themselves are motivated by the theorems that we have previously decided ought to be correct.

Instead of focusing on this strange circularity, philosophers have pretended it does not exist, as if the axiomatic method, proceeding linearly from definition to theorem, were endowed with definitiveness. This is, as every mathematician knows, a subtle fakery to be debunked.

Perform the following thought experiment. Suppose you are given two formal presentations of the same mathematical theory. The definitions of the first presentation are the theorems of the second, and vice versa. This situation frequently occurs in mathematics. Which of the two presentations makes the theory "true?" Neither, evidently: what we have are two presentations of the *same* theory.

This thought experiment shows that mathematical truth is not brought into being by a formal presentation; instead, formal presentation is only a technique for displaying mathematical truth. The truth of a mathematical theory is distinct from the correctness of any axiomatic method that may be chosen for the presentation of the theory.

Mathematizing philosophers have missed this distinction.

The Appeal to Psychology

What will happen to the philosopher who insists on precise statements and clear definitions? Realizing after futile trials that philosophy resists such a treatment, the philosopher will proclaim that most problems previously thought to belong to philosophy are henceforth to be excluded from consideration. He will claim that they are "meaningless," or at best, can be settled by an analysis of their statements that will eventually show them to be vacuous.

This is not an exaggeration. The classical problems of philosophy have become forbidden topics in many philosophy departments. The mere mention of one such problem by a graduate student or by a junior colleague will result in raised eyebrows followed by severe penalties. In this dictatorial regime we have witnessed the shrinking of philosophical activity to an impoverished *problématique*, mainly dealing with language.

In order to justify their neglect of most of the old and substantial questions of philosophy, our mathematizing philosophers have resorted to the ruse of claiming that many questions formerly thought to be philosophical are instead "purely psychological" and that they should be dealt with in the psychology department.

If the psychology department of any university were to consider only one tenth of the problems that philosophers are palming off on them, then psychology would without question be the most fascinating of all subjects. Maybe it is. But the fact is that psychologists have no intention of dealing with problems abandoned by philosophers who have been derelict in their duties.

One cannot do away with problems by decree. The classical problems of philosophy are now coming back with a vengeance in the forefront of science.

Experimental psychology, neurophysiology and computer science may turn out to be the best friends of traditional philosophy. The awesome complexities of the phenomena that are being studied in these sciences have convinced scientists (well in advance of the philosophical establishment) that progress in science will depend on philosophical research in the most classical vein.

The Reductionist Concept of Mind

What does a mathematician do when working on a mathematical problem? An adequate description of the project of solving a mathematical problem might require a thick volume. We will be content with recalling an old saying, probably going back to the mathematician George Pólya: "Few mathematical problems are ever solved directly."

Every mathematician will agree that an important step in solving a mathematical problem, perhaps *the* most important step, consists of analyzing other attempts, either those attempts that have been previously carried out or attempts that he imagines might have been carried out, with a view to discovering how such "previous" approaches failed. In short, no mathematician will ever dream of attacking a substantial mathematical problem without first becoming acquainted with the *history* of the problem, be it the real history or an ideal history reconstructed by the gifted mathematician. The solution of a mathematical problem goes hand in hand with the discovery of the inadequacy of previous attempts, with the enthusiasm that sees through and gradually does away with layers of irrelevancies which formerly clouded the real nature of the problem. In philosophical terms, a mathematician who solves a problem cannot avoid facing up to the *historicity* of the problem. Mathematics is nothing if not a historical subject *par excellence.*

Every philosopher since Heraclitus with striking uniformity has stressed the lesson that all thought is constitutively historical. Until, that is, our mathematizing philosophers came along, claiming that the mind is nothing but a complex thinking machine, not to be polluted by the inconclusive ramblings of bygone ages. Historical thought was dealt a *coup de grâce* by those who today occupy some of the chairs of our philosophy departments. Graduate school requirements in the history of philosophy were dropped, together with language requirements, and in their place we find required courses in mathematical logic.

It is important to uncover the myth that underlies such drastic revision of the concept of mind, that is, the myth that the mind is some sort of mechanical device. This myth has been repeatedly and successfully attacked by the best philosophers of our century (Husserl, John Dewey, Wittgenstein, Austin, Ryle, Croce, to name a few).

According to this myth, the process of reasoning functions like a vending machine which, by setting into motion a complex mechanism reminiscent of Charlie Chaplin's *Modern Times*, grinds out solutions to problems. Believers in the theory of the mind as a vending machine will rate human beings by "degrees" of intelligence, the more intelligent ones being those endowed with bigger and better gears in their brains, as may of course be verified by administering cleverly devised I.Q. tests.

Philosophers believing in the mechanistic myth assert that the solution of a problem is arrived at in just one way: by thinking hard about it. They will go so far as to assert that acquaintance with previous contributions to a problem may bias the well-geared mind. A blank mind, they insist, is better geared to complete the solution process than an informed mind.

This outrageous proposition originates from a misconception of the working habits of mathematicians. Our mathematizing philosophers are failed mathematicians. They gape at the spectacle of mathematicians at work in wide-eyed admiration. To them, mathematicians are superminds who spew out solutions of one problem after another by dint of pure brain power, simply by staring long enough at a blank piece of paper.

The myth of the vending machine that grinds out solutions may appropriately describe the way to solve the linguistic puzzles of today's impoverished philosophy, but this myth is wide of the mark in describing the work of mathematicians, or any kind of serious work.

The fundamental error is an instance of reductionism. The *process* by which the mind works, which may be of interest to physicians but is of no help to working mathematicians, is confused with the *progress* of thought that is required in the solution of any problem. This catastrophic misunderstanding of the concept of mind is the heritage of one hundred-odd years of pseudo-mathematization of philosophy.

The Illusion of Definitiveness

The results of mathematics are definitive. No one will ever improve on a sorting algorithm which has been proved best possible. No one will ever discover a new finite simple group, now that the list has been drawn after a century of research. Mathematics is forever.

We could order the sciences by how close their results come to being definitive. At the top of the list we would find sciences of lesser philosophical interest, such as mechanics, organic chemistry, botany. At the bottom of the list we would find more philosophically inclined sciences such as cosmology and evolutionary biology.

The problems of philosophy, such as mind and matter, reality, perception, are the least likely to have "solutions." We would be hard put to spell out what kind of argument might be acceptable as a "solution to a problem of philosophy." The idea of a "solution" is borrowed from mathematics and tacitly presupposes an analogy between problems of philosophy and problems of science that is fatally misleading.

Philosophers of our day go one step further in their mis-analogies between philosophy and mathematics. Driven by a misplaced belief in definitiveness measured in terms of problems solved, and realizing the futility of any program that promises definitive solutions, they have been compelled to get rid of all classical problems. And where do they think they have found problems worthy of them? Why, in the world of facts!

Science deals with facts. Whatever traditional philosophy deals with, it is not facts in any known sense of the word. Therefore, traditional philosophy is meaningless.

This syllogism, wrong on several counts, is predicated on the assumption that no statement is of any value unless it is a statement of fact. Instead of realizing the absurdity of this vulgar assumption, philosophers have swallowed it, hook, line and sinker, and have busied themselves in making their living on facts.

But philosophy has never been equipped to deal directly with facts, and no classical philosopher has ever considered facts to be any of his business. Nobody will ever turn to philosophy to learn facts. Facts are the business of science, not of philosophy.

And so, a new slogan had to be coined: philosophy *should* be dealing with facts.

This "should" comes at the end of a long normative line of "shoulds." Philosophy should be precise; it should follow the rules of mathematical logic; it should define its terms carefully; it should ignore the lessons of the past; it should be successful at solving its problems; it should produce definitive solutions.

"Pigs should fly," the old saying goes.

But what is the standing of such "shoulds," flatly negated as they are by two thousand years of philosophy? Are we to believe the not so subtle insinuation that the royal road to right reasoning will at last be ours if we follow these imperatives?

There is a more plausible explanation of this barrage of "shoulds." The reality we live in is constituted by a myriad contradictions, which traditional philosophy has taken pains to describe with courageous realism. But contradiction cannot be confronted by minds who have put all their eggs in the basket of precision and definitiveness. The real world is filled with absences, absurdities, abnormalities, aberrances, abominations, abuses, with *Abgrund*. But our latter-day philosophers are not concerned with facing up to these discomforting features of the world, nor to any relevant features whatsoever. They would rather tell us what the world *should* be like. They find it safer to escape from distasteful description of what is into pointless prescription of what isn't. Like ostriches with their heads buried in the sand, they will meet the fate of those who refuse to remember the past and fail to face the challenges of our difficult present: increasing irrelevance followed by eventual extinction.

CHAPTER VIII

Philosophy and Computer Science

In all civilizations of all times, astrologers have been relied on to predict the future. Not too long ago, they were rudely displaced by scientists. Nowadays, the scientists' services in futurology are widely sought by businesses and governments, and their expertise is amply rewarded. Scientists miss no occasion to regale the world with ponderous descriptions about the shape of things to come.

I will not disappoint you by neglecting this glorious calling of ours. But I hasten to say that the future becomes more predictable once we realize that what is happening today and what will happen tomorrow is not the random stroke of fate. The future can be read in the past creations of artists, poets, philosophers, and scientists. We recognize today's surroundings in the canvases of Cézanne and Van Gogh because their imagination contributed to the shaping of today's landscapes. We benefit from reading the *Duino Elegies* and *Ossi di Seppia*[1] because the poets Rilke and Montale invented the meaning life has today. Our grandchildren will recognize their own world in reading the parodies of Donald Barthelme or the plays of Samuel Beckett, which we find baffling and scary.

Classical mechanics is now seen as a limiting case of quantum mechanics because the way was paved for us by the genial insights of Schrödinger, Heisenberg and Fermi. Our children find themselves at ease with computer languages because decades ago Peano, Gödel and Church braved the hostility of the mathematical establishment of their time by developing symbolic logic.

The daily images of our time are not our invention: they were invented by some gifted mind long before we were born. The revolutions we think are currently happening are the remnants of the forgotten theories of some long-dead philosopher.

One such revolution, the revolution that is being brought about by the computer and by the new sciences of mind, is deeply indebted to the work of the great philosophers of the first half of this century.

A new concept of mind is spearheaded by the notion of a computer program. Everyone today realizes that a computer program is not an object, as the word "object" is understood in ordinary speech when referring to physical objects. Everyone today realizes that a mere description of the hardware necessary for a computer program to work misses the *concept* of a computer program. *The same* computer program may be transferred from one type of hardware to another. The computer program is endowed with an *identity* of its own, an identity that coincides with the *function* the program is expected to perform over and over. Such an item is independent of whatever hardware may be required for the working of the program. A computer scientist will never confuse the item that is called "program" with the underlying hardware that makes it work. In philosophical terms, such confusion is an instance of the philosophical error known as reduction.

The computer scientist is led by thought experiments on computer programming to a fundamental realization: a computer program's identity is an item that is related to the hardware by a peculiar relation philosophers call *Fundierung*. The computer scientist has become a convinced antireductionist. This argument was anticipated long ago by Edmund Husserl.

Drawing on the phenomenological description of computer programs, the brain scientist will try to find an analogous argument based on the *Fundierung* relation in dealing with the more difficult case of the brain. Here the spadework of the three great philosophers of this century— Husserl, Heidegger, Wittgenstein— proves to be of inestimable value. The assertion that all understanding is a *process* that goes on *in* the mind is revealed to be another instance of reductionism, much like the identification of hardware with software. Have we ever analyzed what is meant by the word "process" in dealing with mental phenomena? Have we paused to reflect on the physical notion of containment that is implicitly assumed when the preposition "in" is associated with the mind?

The word "process" tacitly assumes a sequence of events linked together by causal connections. It takes for granted that such a sequence develops in time. The reductionist concept of "process" is borrowed from classical mechanics, our paradigm of science. Are we entitled to extrapolate the conceptual framework of mechanics to phenomena that display no evidence of mechanism? Is there evidence for assuming any temporal linearity in mental phenomena? It may turn out that whatever goes on in the brain is not a process at all in the current sense of this term, and that our extrapolation to cognitive science of the properties of mechanical processes is unwarranted. The conceptual framework required of brain science will be a radically new one. The concepts proper to this new science will bear no relationship to terms from everyday speech, terms like "process" and "time" which we have carelessly used. Are we making an error similar to that of a would-be chemist who falls flat on his face after taking for granted that the basic components of matter should be earth, air, fire and water?

The great philosophers of our century have dealt with these problems with amazing foresight and unmatched philosophical rigor. We ignore their arguments only at the cost of painful repetition and embarrassing rediscovery. The once abstruse conclusions of Husserlian phenomenology, the exasperating counterexamples in Wittgenstein's *Philosophical Investigations* will be vindicated by the demands of intellectual honesty of the computer scientist who has to make programs that work, and by the brain scientist's search for a theory that will adequately fit his experimental data. The theme of our time is the improbable but inevitable marriage of philosophy with the new sciences of the mind.

Even in our days of constantly predicted revolutions, it is difficult not to be led to an optimistic conclusion. The new sciences of the computer and the brain will validate the philosophers' theories. But what is more important, they will achieve a goal that philosophy has been unable to attain. They will deal the death stroke to the age-old prejudices that have beset the concept of mind.

CHAPTER IX

The Phenomenology of Mathematical Truth

Like artists who fail to give an accurate description of how they work, like scientists who believe in unrealistic philosophies of science, mathematicians subscribe to a concept of mathematical truth that runs contrary to the truth.

The arbitrariness of the accounts professionals give of the practice of their professions is too universal a phenomenon to be brushed off as a sociological oddity. We will try to unravel the deep-seated reasons for such a contrast between honest practice and trumped-up theory, omitting considerations of psychological aspects, and focusing instead on a philosophical explanation.

We will take a view of mathematical activity drawn from observed fact in opposition to the normative assertions of certain philosophers of mathematics. An honest concept of mathematical truth must emerge from a dispassionate examination of what mathematicians do, rather than from what mathematicians *say* they do, or from what philosophers think mathematicians *ought* to do.

Truth and Tautology

The accepted description of mathematical truth is roughly the following. A mathematical theory consists of axioms, primitive notions, notation, and rules of inference. A mathematical statement is held to be true if it is correctly derived from the axioms by applications of the rules of inference. A mathematical theory will include all possible true statements that can be derived from the axioms.

The truth of the theorems of a mathematical theory is seldom seen by merely staring at the axioms, as anyone who has ever worked in mathematics will tell you. Nevertheless we go on believing that the truth of all theorems can "in principle" be "found" in the axioms. Terms like "in principle" and "found" are frequently used to denote the "relationship" of the theorems of mathematics with the axioms from which they are derived, and the meaning of such terms as "found," "relationship," and "in principle," is glibly taken for granted. Discussions of the conditions of possibility of such "relationships" are downplayed; what matters is arriving by the shortest route at the expected conclusion, decided upon in advance, which will be the peremptory assertion that all mathematical truths are "ultimately" tautological.

No one will go so far as to confuse tautology with triviality. The theorems of mathematics may be "ultimately" or "in principle" tautological, but such tautologies more often than not require strenuous effort to be proved. Thus, although the theorems of mathematics may be tautological consequences of the axioms, at least "in principle," nonetheless, such tautologies are neither immediate nor evident.

But what is the point of using the word "tautology" in this discussion? Of what help to the understanding of mathematics is it to assert that mathematical theorems are "in principle" tautological? Such an assertion, far from clearing the air, is a way of discharging the burden of understanding what mathematical truth is onto the catchall expression "in principle."

What happens is that the expression "in principle" is carefully cooked up to conceal a normative term, namely, the word "should." We say that mathematical theorems are "ultimately," "in principle," "basically" tautological, when we mean to say that such theorems *should* be evident from the axioms, that the intricate succession of syllogistic inferences by which we prove a theorem, or by which we understand someone else's theorem, is only a temporary prop that *should*, sooner or later, *ultimately* let us see the conclusion as an inevitable consequence of the axioms.

The role of the implicit normative term "should" in what appears to be no more than a description has seldom been brought out into the open. Perhaps it will turn out to be related to the sense in which the word "truth" is used in the practice of mathematics.

Formal Truth

We will establish a distinction between the concept of truth as it is used by mathematicians and another, superficially similar concept, also unfortunately denoted by the word "truth," which is used in mathematical logic and has been widely adopted in analytic philosophy.

The narrow concept of truth coming to us from logic may more properly be called "formal verification." The logician denotes by the word "truth" either the carried-out formal verification of the correctness of derivation of a statement from axioms, or else its semantic counterpart, the carried-out formal verification that a statement holds in all models.

This concept of truth is a derived one, that is, it already presupposes another concept of truth, the one by which mathematicians tacitly abide. To realize that the concept of formal verification is hopelessly wide of the mark, perform the following eidetic variation. Imagine a mathematics lesson. Can you envisage a mathematics teacher training his or her pupils *exclusively* in their ability to produce logically impeccable proofs from axioms? Or even (to take a deliberately absurd exaggeration) a teacher whose main concern is his students' efficiency in spotting the consequences of a "given" set of axioms?

In the formal concept of truth, axioms are not questioned, and proving theorems is viewed as a game in which all pretense of telling the truth is suspended. But no self-respecting teacher of mathematics can afford to pawn off on a class the axioms of a theory without giving some motivation, nor can the class be expected to accept the results of the theory (the theorems) without some sort of justification other than formal verification. Far from taking axioms for granted and beginning to spin off their consequences, a teacher who wants to be understood will engage in a back-and-forth game in which the axioms are "justified" by the strength of the theorems they are capable of proving; and, once the stage is set, the theorems themselves, whose truth has previously been made intuitively evident, are finally made inevitable by formal proofs which come almost as an afterthought, as the last bit of crowning evidence of a theory that has already been made plausible by nonformal, nondeductive, and at times even nonrational discourse.

A good teacher who is asked to teach the Euler-Schläfli-Poincaré formula giving the invariance of the alternating sum of the number of sides of a polyhedron will acknowledge to his class that such a formula was believed to be true long before any correct definition of a polyhedron was known, and will not hide the fact that the verification of the formula in a formal topological setting is an afterthought. The teacher will insist that the class not confuse such formal verification with the factual/worldly truth of the assertion. It is such a factual/worldly truth that motivates all formal presentations, not vice versa, as formalist philosophers of mathematics pretend is the case.

Thus, what matters to any teacher of mathematics is the teaching of what mathematicians in their shoptalk informally refer to as the "truth" of a theory, a truth that has to do with the concordance of a statement with the facts of the world, like the truth of any physical law. In the teaching of mathematics, the truth that is demanded by the students and provided by the teacher is such a factual/worldly truth, and not the formal truth that one associates with the game of theorem-proving. A good teacher of mathematics is one who knows how to disclose the full light of such factual/worldly truth to students while training them in the skills of carefully *recording* such truth. In our age such skills happen to coincide with giving an accurate formal-deductive presentation.

Despite such overwhelming evidence from the practice of mathematics, there are influential thinkers who claim that discussions of factual/worldly truth should be permanently confined to the slums of psychology. It is more comfortable to deal with notions of truth that are prescribed in a *pensée de survol* which save us from direct commerce with mathematicians.

The preference for the tidy notion of formal verification, rather than the unkempt notion of truth that is found in the real world of mathematicians, has an emotional source. What has happened is that the methods that have so brilliantly succeeded in the definition and analysis of formal systems have failed in the task of giving an account of other equally relevant features of the mathematical enterprise. Philosophers of mathematics display an irrepressible desire to tell us as quickly as possible what mathematical truth *ought* to be while bypassing the descriptive legwork that is required for an accurate account of the truth by which mathematicians live.

All formalist theories of truth are reductionistic. They derive from an unwarranted identification of mathematics with the axiomatic method of presentation of mathematics. The fact that there are only five regular solids in three-dimensional Euclidean space may be presented in widely differing axiomatic settings; no one doubts their truth, regardless of the axiomatics chosen for their justification. Such a conclusive example should be proof enough that the relation between the truth of mathematics and the axiomatic truth which is indispensable in the presentation of mathematics is a relation of *Fundierung*.

Nevertheless, in our time the temptation of psychologism is again rearing its head, and we find ourselves forced to revive old and elementary antireductionist precautions. No state of affairs can be "purely psychological." The psychological aspects of the teaching of mathematics must of necessity hark back to a worldly truth of mathematics. Viewing the truth that the mathematics teacher appeals to as a mere psychological device is tantamount to presupposing a worldly mathematical truth while refusing to thematize it. *Nihil est in intellectu quod prius non fierit in mundo*, we could say, murdering an old slogan.

Mathematical truth is philosophically no different from the truth of physics or chemistry. Mathematical truth results from the formulation of facts that are out there in the world, facts that are independent of our whim or of the vagaries of axiomatic systems.

Truth and Triviality

We will now argue *against* the thesis we have just stated, again by taking some events in the history of mathematics as a guide. We will give an example that displays how observation of the practice of mathematics leads to a more sophisticated concept of truth than the one we have stated.

Our example is the history of the prime number theorem. This result was conjectured by Gauss after extensive numerical experimentation, guided by a genial intuition. No one seriously doubted the truth of the theorem after Gauss had conjectured it and verified it numerically. However, mathematicians cannot afford to behave like physicists who take experimental verification as confirmation of the truth. Thanks to high-speed computers we know what Gauss could only guess, that certain conjectures in number theory may fail for integers so large as to lie beyond the reach of even the best of today's computers, and thus, that formal proof is more indispensable than ever.

The proof of the prime number theorem was obtained simultaneously and independently around the turn of the century by the mathematicians Hadamard and de la Vallée Poussin. Both proofs, which were very similar, relied on the latest techniques in the theory of functions of a complex variable. They were justly hailed as a great event in the history of mathematics. However, to the best of my knowledge, no one verbalized at the time the innermost reason for the mathematician's glee. An abstruse theory which was at the time the cutting edge of mathematics, namely, the theory of functions of a complex variable, developed in response to geometric and analytic problems, turned out to be the key to settling a conjecture in number theory, an entirely different field.

The mystery as well as the glory of mathematics lies not so much in the fact that abstract theories turn out to be useful in solving problems but in the fact that — wonder of wonders— a theory meant for one type of problem is often the only way of solving problems of entirely different kinds, problems for which the theory was not intended. These coincidences occur so frequently that they must belong to the essence of mathematics. No philosophy of mathematics shall be excused from explaining such occurrences.

One might think that once the prime number theorem was proved other attempts at proving it by altogether different techniques would be abandoned as fruitless. But this is not what happened after Hadamard and de la Vallée Poussin. Instead, for about fifty years thereafter, paper after paper began to appear in the best mathematics journals that provided nuances, simplifications, alternative routes, slight generalizations, and eventually alternative proofs of the prime number theorem. For example, in the thirties, the American mathematician Norbert Wiener developed an extensive theory of Tauberian theorems that unified a great number of disparate results in classical mathematical analysis. The outstanding application of Wiener's theory, widely acclaimed throughout the mathematical world, was precisely a new proof of the prime number theorem.

Confronted with this episode in mathematical history, an outsider might be led to ask, "What? A theory lays claim to be a novel contribution to mathematics by proudly displaying as its main application a result that by that time had been cooked in several sauces? Isn't mathematics supposed to solve *new* problems?"

To allay suspicions, let it be added that Wiener's main Tauberian theorem was and still is viewed as a great achievement. Wiener was the first to succeed in injecting an inkling of purely conceptual insight into a proof that had heretofore appeared mysterious. The original proof of the prime number theorem mysteriously related the asymptotic distribution of primes to the behavior of the zeroes of a meromorphic function, the Riemann zeta function. Although this connection had been established some time ago by Riemann, and although the logic of such an improbable connection had been soundly tested by several mathematicians who had paved the way for the first proof, nevertheless, the proof could not be said to be based upon obvious and intuitive concepts. Wiener succeeded in showing, by an entirely different route than the one followed by his predecessors, and one that was just as unexpected, that there might be a conceptual underpinning to the distribution of primes.

Wiener's proof had a galvanizing effect. From that time on, it was believed that the proof of the prime number theorem could be made elementary.

What does it mean to say that a proof is "elementary?" In the case of the prime number theorem, it means that an argument is given that shows the "analytic inevitability" (in the Kantian sense of the expression) of the prime number theorem on the basis of an analysis of the concept of prime without appealing to extraneous techniques.

It took another ten years and a few hundred research papers to remove a farrago of irrelevancies from Wiener's proof. The first elementary proof of the prime number theorem, one that "in principle" used only elementary estimates of the relative magnitudes of primes was finally obtained by the mathematicians Erdös and Selberg. Again, their proof was hailed as a new milestone in number theory.

Elementary proofs are seldom simple. Erdös and Selberg's proof added up to a good fifty pages of elementary but thick reasoning and was longer and harder to follow than any of the preceding ones. It did, however, have the merit of relying only upon notions that were "intrinsic" to the definition of prime number, as well as on a few other elementary facts going back to Euclid and Eratosthenes. In principle, their proof showed how the prime number theorem could be boiled down to a fairly trivial argument once the basic notions had been properly grasped. But only in principle. It took another few hundred research papers, whittling down Erdös and Selberg's argument to the barest core, until, in the middle sixties, the American mathematician

Norman Levinson (who was Norbert Wiener's research student) published a short note bearing the title "An elementary proof of the prime number theorem."[1] Despite its modest title, Levinson's note outlined a purely elementary proof of the prime number theorem, one that can be followed by careful reading by anyone with no more knowledge of mathematics than that of undergraduates at an average American college.

After Levinson's paper, research on proofs of the prime number theorem dwindled. Levinson's proof of the prime number theorem, or one of several variants that have been discovered since, is now part of the curriculum of an undergraduate course in number theory.

What philosophical conclusions can we draw from this fragment of mathematical history?

Leafing through any of the three thousand-odd journals that publish original mathematical research, one soon discovers that few published research papers present solutions of as yet unsolved problems; fewer still are formulations of new theories. The overwhelming majority of research papers in mathematics is concerned not with proving, but with reproving; not with axiomatizing, but with re-axiomatizing; not with inventing, but with unifying and streamlining; in short, with what Thomas Kuhn calls "tidying up."

In the face of this evidence, we are forced to choose between two conclusions. The first is that the quality of mathematical research in our time is lower than we were led to expect. But what kind of evidence could have led us to such expectations? Certainly not the history of mathematics in the eighteenth or nineteenth centuries. Publication in mathematics in these centuries followed the same pattern we have described, as a dispassionate examination of the past will show.

Only one other conclusion is possible. Our preconceived ideas of what mathematical research *should* be do not correspond to the reality of mathematical research. The mathematician's profession does not consist of inventing imaginative new theories out of nowhere that will unravel the mysteries of nature.

We hasten to add that the opposite opinion of the mathematician's activity is equally wrong: it is not true that mathematicians do *not* invent new theories. They do; in fact, they make their living by doing so. However, most research papers in mathematics are not easily classified with respect to their originality. A stern judge would classify all mathematics papers as unoriginal, except perhaps two or three in a century; a more liberal one would find a redeeming spark of originality in most papers that appear in print. Even papers that purport to give solutions of heretofore unsolved problems can be severely branded as either exercises of varying degrees of difficulty, or more kindly as pathways opening up brave new worlds.

The value of a mathematics research paper is not deterministically given, and we err in forcing our judgment about the worth of a paper to conform to objectivistic standards. Papers which only twenty years ago were taken as fundamental are nowadays viewed as misguided.

Newer theories do not subsume and expand on preceding theories by quantitative increases in the amount of information. The invention of theories and the solution of difficult problems are not processes evolving linearly in time. Never more than today has the oversimplification of linearity been off the mark. We are witnessing a return to the concrete mathematics of the nineteenth century after a long period of abstraction; algorithms and techniques that were once made fun of are now being revalued after a century of interruption. Contemporary mathematics, with its lack of a unifying trend, its historical discontinuities, and its lapses into the past, provides further evidence of the end of two embarrassing Victorian heritages: the idea of progress and the myth of definitiveness.

The Ambiguity of Truth

We have sketched two seemingly clashing concepts of mathematical truth. Both concepts force themselves upon us when we observe the development of mathematics.

The first concept is similar to the classical concept of the truth of a law of natural science. According to this first view, mathematical theorems are statements of fact; like all facts of science, they are discovered by observation and experimentation. The philosophical theory of mathematical facts is therefore not essentially distinct from the theory of any other scientific facts, except in phenomenological details. For example, mathematical facts exhibit greater precision when compared to the facts of certain other sciences, such as biology. It matters little that the facts of mathematics might be "ideal," while the laws of nature are "real," as philosophers used to say some fifty years ago. Whether real or ideal, the facts of mathematics are out there in the world and are not creations of someone's mind. Both mathematics and natural science have set themselves the same task of discovering the regularities in the world. That some portions of this world may be real and others ideal is a remark of little relevance.

The second view seems to lead to the opposite conclusion. Proofs of mathematical theorems, such as the proof of the prime number theorem, are achieved at the cost of great intellectual effort. They are then gradually whittled down to trivialities. Doesn't this temporal process of simplification that transforms a fifty-page proof into a half-page argument betoken the assertion that the theorems of mathematics are creations of our own intellect? Doesn't it follow from these observations that the original difficulty of a mathematical theorem, the difficulty with which we struggle when we delude ourselves about having "discovered" a new theorem, is really due to human frailty alone, a frailty that some stronger mind will at some later date dissipate by displaying a triviality that we had failed to acknowledge?

Every mathematical theorem is eventually proved trivial. The mathematician's ideal of truth is triviality, and the community of mathematicians will not cease its beaver-like work on a newly discovered result until it has shown to everyone's satisfaction that all difficulties in the early proofs were spurious, and only an analytic triviality is to be found at the end of the road. Isn't the progress of mathematics — if we can speak of progress at all — just a gradual awakening from "*el sueño de la razón*?"

The way I propose out of the paradox of these two seemingly irreconcilable views of mathematics will use an argument from Edmund Husserl. The same argument is useful in a large number of other philosophical puzzles and justly deserves a name of its own. I should like to baptize it "the *ex universali* argument."

Let us summarize the problem at hand. On the one hand, mathematics is undoubtedly the recording of phenomena which are not arbitrarily determined by the human mind. Mathematical facts follow the unpredictable *a posteriori* behavior of a nature which is, in Einstein's words, *raffiniert* but not *boshaft*. On the other hand, the power of reason sooner or later reduces every such fact to an analytic statement amounting to a triviality. How can both these statements be true?

But observe that such flagrantly duplicitous behavior is not the preserve of mathematical research alone. The facts of other sciences, physics and chemistry, and someday (we firmly believe) even biology, exhibit the same duplicitous behavior.

Any law of physics, when finally ensconced in a proper mathematical setting, turns into a mathematical triviality. The search for a universal law of matter, a search in which physics has been engaged throughout this century, is actually the search for a trivializing principle, for a universal "nothing but." The unification of chemistry that has been wrought by quantum mechanics is not differently motivated. And the current fashion of molecular biology can be attributed to the glimmer of hope that this glamorous new field is offering biology, for the first time in the history of the life sciences, to finally escape from the whimsicality of natural randomness into the coziness of Kantian analyticity.

The ideal of *all* science, not only of mathematics, is to do away with any kind of synthetic *a posteriori* statement and to leave only analytic trivialities in its wake. Science may be defined as the transformation of synthetic facts of nature into analytic statements of reason.

Thus, the *ex universali* argument shows that the paradox that we believed to have discovered in the truth of mathematics is shared by the truth of all the sciences. Although this admission does not offer immediate solace to our toils, it is nevertheless a relief to know that we are not alone in our misery. More to the point, the realization of the universality of our paradox exempts us from even attempting to reason ourselves out of it within the narrow confines of the philosophy of mathematics. At this point, all we can do is turn the whole problem over to the epistemologist or to the metaphysician and remind them of the old injunction: *hic Rhodus, hic salta*.

CHAPTER X

The Phenomenology of Mathematical Beauty

Whereas painters and musicians are likely to be embarrassed by references to the beauty of their work, mathematicians enjoy discussions of the beauty of mathematics. Professional artists stress the technical rather than the aesthetic aspects of their work. Mathematicians, instead, are fond of passing judgment on the beauty of their favored pieces of mathematics. A cursory observation shows that the characteristics of mathematical beauty are at variance with those of artistic beauty. Courses in "art appreciation" are fairly common; it is unthinkable to find any "mathematical beauty appreciation" courses. We will try to uncover the sense of the term "beauty" as it is used by mathematicians.

What Kind of Mathematics Can Be Beautiful?

Theorems, proofs, entire mathematical theories, a short step in the proof of some theorem, and definitions are at various times thought to be beautiful or ugly by mathematicians. Most frequently, the word "beautiful" is applied to theorems. In the second place we find proofs; a proof that is deemed beautiful tends to be short. Beautiful theories are also thought of as short, self-contained chapters fitting within broader theories. There are complex theories which every mathematician agrees to be beautiful, but these examples are not the ones that come to mind in making a list of beautiful pieces of mathematics. Theories that mathematicians consider to be beautiful seldom agree with the mathematics thought to be beautiful by the educated public. For example, classical Euclidean geometry is often proposed by nonmathematicians as a paradigm of a beautiful mathematical theory, but I have not heard it classified as such by professional mathematicians.

It is not uncommon for a definition to seem beautiful, especially when it is new. However, mathematicians are reluctant to admit the beauty of a definition; it would be interesting to investigate the reasons for this reluctance. Even when not explicitly acknowledged as such, beautiful definitions give themselves away by the success they meet. A peculiarity of twentieth century mathematics is the appearance of theories where the definitions far exceed the theorems in beauty.

The most common instance of beauty in mathematics is a brilliant step in an otherwise undistinguished proof. Every budding mathematician quickly becomes familiar with this instance of mathematical beauty.

These instances of mathematical beauty are often independent of each other. A beautiful theorem may not be blessed with an equally beautiful proof; beautiful theorems with ugly proofs frequently occur. When a beautiful theorem is missing a beautiful proof, attempts are made by mathematicians to provide new proofs that will match the beauty of the theorem, with varying success. It is however impossible to find beautiful proofs of theorems that are not beautiful.

Examples

The theorem stating that in three dimensions there are only five regular solids (the Platonic solids) is generally considered to be beautiful; none of the proofs of this theorem, however, at least none of those that are known to me, can be said to be beautiful. Similarly, the prime number theorem is a beautiful result regarding the distribution of primes, but none of its proofs can be said to be particularly beautiful.

Hardy's opinion that much of the beauty of a mathematical statement or of a mathematical proof depends on the element of surprise is, in my opinion, mistaken.[1] True, the beauty of a piece of mathematics is often perceived with a feeling of pleasant surprise; nonetheless, one can find instances of surprising results which no one has ever thought of classifying as beautiful. Morley's theorem, stating that the adjacent trisectors of an arbitrary triangle meet in an equilateral triangle is unquestionably surprising, but neither the statement nor any of the proofs are beautiful despite repeated attempts to provide streamlined proofs. A great many theorems of mathematics, when first published, appear to be surprising; some twenty years ago the proof of the existence of non-equivalent differentiable structures on spheres of high dimension was thought to be surprising but it did not occur to anyone to call such a fact beautiful, then or now.

Instances of theorems that are both beautiful and surprising abound. Often such surprise results from a proof that borrows ideas from another branch of mathematics. An example is the proof of the Weierstrass approximation theorem that uses the law of large numbers of probability .

An example of mathematical beauty upon which all mathematicians agree is Picard's theorem, asserting that an entire function of a complex variable takes all values with at most one exception. The limpid statement of this theorem is matched by the beauty of the five-line proof provided by Picard .

Axiom systems can be beautiful. Church's axiomatization of the propositional calculus, which is a simplified version of the one previously given by Russell and Whitehead in *Principia Mathematica*, is quite beautiful. Certain re-elaborations of the axioms of Euclidean geometry that issue from Hilbert's *Foundations of Geometry*[2] are beautiful (for example, Coxeter's). Hilbert's original axioms were clumsy and heavy-handed, and required streamlining; this was done by several mathematicians of the last hundred years.

The axiomatization of the notion of category, discovered by Eilenberg and Mac Lane in the forties, is an example of beauty in a definition, though a controversial one. It has given rise to a new field, category theory, which is rich in beautiful and insightful definitions and poor in elegant proofs. The basic notions of this field, such as adjoint and representable functor, derived category, and topos, have carried the day with their beauty, and their beauty has been influential in steering the course of mathematics in the latter part of this century; however, the same cannot be said of the theorems which remain clumsy.

An example of a beautiful theory on which most mathematicians are likely to agree is the theory of finite fields, initiated by E. H. Moore. Another is the Galois theory of equations, which invokes the once improbable notion of a group of permutations in proving the unsolvability by radicals of equations of degree greater than four. The beauty of this theory has inspired a great many expositions. It is my opinion that so far they have failed to convey the full beauty of the theory, even the renowned treatise written by Emil Artin in the forties.[3]

This example shows that the beauty of a mathematical theory is independent of the aesthetic qualities, or the lack of them, of the theory's rigorous expositions. Some beautiful theories may never be given a presentation that matches their beauty. Another such instance of a beautiful theory which has never been matched in beauty of presentation is Gentzen's natural deduction.

Instances of profound mathematical theories in which mathematical beauty plays a minor role abound. The theory of differential equations, both ordinary and partial, is fraught with ugly theorems and awkward arguments. Nonetheless, the theory has exerted a strong fascination on a large number of mathematicians, aside from its applications.

Instances can also be found of mediocre theories of questionable beauty which are given brilliant, exciting presentations. The mixed blessing of an elegant presentation will endow the theory with an ephemeral beauty which seldom lasts beyond the span of a generation or a school of mathematics. The theory of the Lebesgue integral, viewed from the vantage point of one hundred years of functional analysis, has received more elegant presentations than it deserves. Synthetic projective geometry in the plane held great sway between 1850 and 1940. It is an instance of a theory whose beauty was largely in the eyes of its beholders. Numerous expositions were written of this theory by English and Italian mathematicians (the definitive one being the one given by the American mathematicians Veblen and Young). These expositions vied with one another in elegance of presentation and in cleverness of proof; the subject became required by universities in several countries. In retrospect, one wonders what all the fuss was about. Nowadays, synthetic geometry is largely cultivated by historians, and an average mathematician ignores the main results of this once flourishing branch of mathematics. The claim that has been raised by defenders of synthetic geometry, that synthetic proofs are more beautiful than analytic proofs, is demonstrably false. Even in the nineteenth century, invariant-theoretic techniques were available that could have provided elegant, coordinate-free analytic proofs of geometric facts without resorting to the gymnastics of synthetic reasoning and without having to stoop to using coordinates.

Beautiful presentations of an entire mathematical theory are rare. When they occur, they have a profound influence. Hilbert's *Zahlbericht*,[4] Weber's *Algebra*,[5] Feller's treatise on probability, certain volumes of Bourbaki, have had influence on the mathematics of our day; one rereads these books with pleasure, even when familiar with their content. Such high-caliber expository work is more exploited than rewarded by the mathematical community.

Finally, it is easy to produce examples of a particular step in a theorem which is generally thought to be beautiful. In the theory of noncommutative rings, the use of Schur's lemma has often been thought of as a beautiful step. The application of the calculus of residues in the spectral theory of linear operators in Hilbert space is another such instance. In universal algebra, the two-sided characterization of a free algebra in the proof of Birkhoff's theorem on varieties is yet another such instance.

The Objectivity of Mathematical Beauty

The rise and fall of synthetic geometry shows that the beauty of a piece of mathematics is dependent upon schools and periods. A theorem that is in one context thought to be beautiful may in a different context appear trivial. Desargues' theorem is beautiful when viewed as a statement of synthetic projective geometry but loses all interest when stated in terms of coordinates.

Many occurrences of mathematical beauty fade or fall into triviality as mathematics progresses. However, given the historical period and the context, one finds substantial agreement among mathematicians as to which mathematics is to be regarded as beautiful. This agreement is not merely the perception of an aesthetic quality which is superimposed on the content of a piece of mathematics. A piece of mathematics that is agreed to be beautiful is more likely to be included in school curricula; the discoverer of a beautiful theorem is rewarded by promotions and awards; a beautiful argument will be imitated. In other words, the beauty of a piece of mathematics does not consist merely of the subjective feelings experienced by an observer. The beauty of a theorem is an objective property on a par with its truth. The truth of a theorem does not differ from its beauty by a greater degree of objectivity.

Mathematical truth is endowed with an absoluteness that few other phenomena can hope to match. On closer inspection, one realizes that this definitiveness needs to be tempered. The dependency of mathematical truth upon proof proves to be its Achilles's heel. A proof that will pass today's standard of rigor may no longer be thought to be rigorous by future generations. The entire theory upon which some theorem depends may at some later date be shown to be incomplete. Standards of rigor and relevance are context-dependent, and any change in these standards leads to a concomitant change in the standing of a seemingly timeless mathematical assertion.

Similar considerations apply to mathematical beauty. Mathematical beauty and mathematical truth share the fundamental property of objectivity, that of being inescapably context-dependent. Mathematical beauty and mathematical truth, like any other objective characteristics of mathematics, are subject to the laws of the real world, on a par with the laws of physics. Context-dependence is the first and basic such law.

A Digression into Bounty Words

A psychologist of my acquaintance received a grant to study how mathematicians work. She decided that creativity plays a crucial role in mathematics. She noticed that an estimate of a mathematician's creativity is made at crucial times in his or her career. By observation of mathematicians at work, she was led to formulate a theory of mathematical creativity, and she devised ways of measuring it. She described how creativity fades in certain individuals at certain times. She outlined ways of enhancing creativity. In her final report, she made the recommendation to her sponsors that mathematics students should, at some time in their careers, be required to register for a course in creativity. Some college presidents took her suggestion seriously and proceeded to hire suitable faculty.

Our friend was seriously in error. It is impossible to deal with mathematical creativity in the way she suggested. It is impossible to measure or teach creativity for the simple reason that creativity is a word devoid of identifiable content. One can characterize a mathematical paper as "creative" only after the paper has been understood. It is, however, impossible to produce on commission a "creatively written" mathematical paper. "Creativity" is what we propose to call a "bounty word," a word that promises some benefit that cannot be controlled or measured, and that can be attained as the unpredictable byproduct of some identifiable concrete activity.

Other bounty words like "creativity" are "happiness," "saintlihood," and "mathematical beauty." Like creativity and happiness, mathematical beauty cannot be taught or sought after; nevertheless, any mathematician will come up with some beautiful statement or some beautiful proof at unpredictable times. The error my lady friend made might be called "the bounty error." It consists of endowing a bounty word with measurable content.

It is unlikely that a mathematician will commit the bounty error in regards to mathematical beauty. Passing judgment on a piece of mathematics on the basis of its beauty is a risky business. In the first place, theorems or proofs which are agreed upon to be beautiful are rare. In the second place, mathematical research does not strive for beauty. Every mathematician knows that beauty cannot be *directly* sought. Mathematicians work to solve problems and to invent theories that will shed new light and not to produce beautiful theorems or pretty proofs.

Even in the teaching of mathematics beauty plays a minor role. One may lead a class to the point where the students will appreciate a beautiful result. However, attempts to arouse interest in the classroom on the basis of the beauty of the material are likely to backfire. Students may be favorably impressed by the elegance of a teacher's presentation, but they can seldom be made aware of beauty. Appreciation of mathematical beauty requires familiarity with a mathematical theory, and such familiarity is arrived at at the cost of time, effort, exercise, and *Sitzfleisch* rather than by training in beauty appreciation.

There is a difference between mathematical beauty and mathematical elegance. Although one cannot strive for mathematical beauty, one can achieve elegance in the presentation of mathematics. In preparing to deliver a mathematics lecture mathematicians often choose to stress elegance, and succeed in recasting the material in a fashion that everyone will agree is elegant. Mathematical elegance has to do with the presentation of mathematics, and only tangentially does it relate to its content. A beautiful proof — for example, Hermann Weyl's proof of the equidistribution theorem — has been presented elegantly and inelegantly. Certain elegant mathematicians have never produced a beautiful theorem.

Mathematical Ugliness

It may help our understanding of mathematical beauty to consider its opposite. The lack of beauty in a piece of mathematics is a frequent occurrence, and it is a motivation for further research. Lack of beauty is related to lack of definitiveness. A beautiful proof is more often than not the definitive proof (though a definitive proof need not be beautiful); a beautiful theorem is not likely to be improved upon, though often it is a motive for the development of definitive theories in which it may be ensconced.

Beauty is seldom associated with pioneering work. The first proof of a difficult theorem is seldom beautiful.

Strangely, mathematicians do not like to admit that much mathematical research consists precisely of polishing and refining statements and proofs of known results. However, a cursory look at any mathematics research journal will confirm this state of affairs.

Mathematicians seldom use the word "ugly." In its place are such disparaging terms as "clumsy," "awkward," "obscure," "redundant," and, in the case of proofs, "technical," "auxiliary," and "pointless." But the most frequent expression of condemnation is the rhetorical question, "What is this good for?"

Observe the weirdness of such a question. Most results in pure mathematics, even the deepest ones, are not "good" for anything. In light of such lack of applications, the occurrence of the disparaging question, "What is this good for?" is baffling. No mathematician who poses this rhetorical question about some mathematical theorem really means to ask for a listing of applications. What, then, is the sense of this question? By analyzing the hidden motivation of the question, "What is this good for?" we will come closer to the hidden sense of mathematical beauty.

The Light Bulb Mistake

The beauty of a piece of mathematics is frequently associated with shortness of statement or of proof. How we wish that all beautiful pieces of mathematics shared the snappy immediacy of Picard's theorem. This wish is rarely fulfilled. A great many beautiful arguments are long-winded and require extensive buildup. Familiarity with a huge amount of background material is the condition for understanding mathematics. A proof is viewed as beautiful only after one is made aware of previous clumsier proofs.

Despite the fact that most proofs are long, despite our need for extensive background, we think back to instances of appreciation of mathematical beauty as if they had been perceived in a moment of bliss, in a sudden flash like a light bulb suddenly being lit. The effort put into understanding the proof, the background material, the difficulties encountered in unraveling an intricate sequence of inferences fade and magically disappear the moment we become aware of the beauty of a theorem. The painful process of learning fades from memory and only the flash of insight remains.

We would *like* mathematical beauty to consist of this flash; mathematical beauty *should* be appreciated with the instantaneousness of a light bulb being lit. However, it would be an error to pretend that the appreciation of mathematical beauty is what we vaingloriously feel it should be, namely, an instantaneous flash. Yet this very denial of the factual truth occurs much too frequently.

The light bulb mistake is often taken as a paradigm in teaching mathematics. Forgetful of our learning pains, we demand that our students display a flash of understanding with every argument we present. Worse yet, we mislead our students by trying to convince them that such flashes of understanding are the core of mathematical appreciation.

Attempts have been made to string together beautiful mathematical results and to present them in books bearing such attractive titles as *The One Hundred Most Beautiful Theorems of Mathematics*. Such anthologies are seldom found on a mathematician's bookshelf.

The beauty of a theorem is best observed when the theorem is presented as the crown jewel within the context of a theory. But when mathematical theorems from disparate areas are strung together and presented as "pearls," they are likely to be appreciated only by those who are already familiar with them.

The Concept of Mathematical Beauty

The clue to our understanding of the hidden sense of mathematical beauty is the light bulb mistake. The stark contrast between the effort required for the appreciation of mathematical beauty and the imaginary view mathematicians cherish of a flash-like perception of beauty is the *Leitfaden* that leads us to discover what mathematical beauty is.

Mathematicians are concerned with the truth. In mathematics, however, there is an ambiguity in the use of the word "truth." This ambiguity can be observed whenever mathematicians claim that beauty is the *raison d'être* of mathematics, or that mathematical beauty is that feature that gives mathematics a unique standing among the sciences. These claims are as old as mathematics, and lead us to suspect that mathematical truth and mathematical beauty may be related.

Mathematical beauty and mathematical truth share one important property. Neither of them admits degrees. Mathematicians are annoyed by the graded truth which they observe in other sciences.

Mathematicians ask the question, "What is this good for?" when they are puzzled by some mathematical assertion, not because they are unable to follow the proof or the applications. Quite the contrary. What happens is that a mathematician has been able to verify its truth in the logical sense of the term, but something is still missing. The mathematician who is baffled and asks the question, "What is this good for?" is missing the *sense* of the statement that has been verified to be true. Verification alone does not give us a clue as to the role of a statement within the theory; it does not explain the *relevance* of the statement. In short, the logical truth of a statement does not *enlighten* us as to the *sense* of the statement. *Enlightenment* not truth is what the mathematician seeks when asking, "What is this good for?" Enlightenment is a feature of mathematics about which very little has been written.

The property of being enlightening is objectively attributed to certain mathematical statements and denied to others. Whether a mathematical statement is enlightening or not may be the subject of discussion among mathematicians. Every teacher of mathematics knows that students will not learn by merely grasping the formal truth of a statement. Students must be given some enlightenment as to the *sense* of the statement, or they will quit. Enlightenment is a quality of mathematical statements that one sometimes gets and sometimes misses, like truth. A mathematical theorem may be enlightening or not, just like it may be true or false.

If the statements of mathematics were formally true but in no way enlightening, mathematics would be a curious game played by weird people. Enlightenment is what keeps the mathematical enterprise alive and what gives mathematics a high standing among scientific disciplines.

Mathematicians seldom explicitly acknowledge the phenomenon of enlightenment for at least two reasons. First, unlike truth, enlightenment is not easily formalized. Second, enlightenment admits degrees: some statements are more enlightening than others. Mathematicians dislike concepts admitting degrees, and will go to any length to deny the logical role of any such concept. Mathematical beauty is the expression mathematicians have invented in order to obliquely admit the phenomenon of enlightenment while avoiding acknowledgment of the fuzziness of this phenomenon. They say that a theorem is beautiful when they mean to say that the theorem is enlightening. We acknowledge a theorem's beauty when we see how the theorem "fits" in its place, how it sheds light around itself, like a *Lichtung*, a clearing in the woods. We say that a proof is beautiful when it gives away the secret of the theorem, when it leads us to perceive the inevitability of the statement being proved. The term "mathematical beauty," together with the light bulb mistake, is a trick mathematicians have devised to avoid facing up to the messy phenomenon of enlightenment. The comfortable one-shot idea of mathematical beauty saves us from having to deal with a concept which comes in degrees. Talk of mathematical beauty is a copout to avoid confronting enlightenment, a copout intended to keep our description of mathematics as close as possible to the description of a mechanism. This copout is one step in a cherished activity of mathematicians, that of building a perfect world immune to the messiness of the ordinary world, a world where what we think *should* be true turns out to *be* true, a world that is free from the disappointments, the ambiguities, the failures of that other world in which we live.

CHAPTER XI

The Phenomenology of Mathematical Proof

What is a Mathematical Proof?

Everybody knows what a mathematical proof is. A proof of a mathematical theorem is a sequence of steps which leads to the desired conclusion. The rules to be followed in this sequence of steps were made explicit when logic was formalized early in this century and they have not changed since. These rules can be used to disprove a putative proof by spotting logical errors; they cannot, however, be used to find the missing proof of a mathematical conjecture.

The expression "correct proof" is redundant. Mathematical proof does not admit degrees. A sequence of steps in an argument is either a proof, or it is meaningless. Heuristic arguments are a common occurrence in the practice of mathematics. However, heuristic arguments do not belong to formal logic. The role of heuristic arguments has not been acknowledged in the philosophy of mathematics despite the crucial role they play in mathematical discovery.

The mathematical notion of proof is strikingly at variance with notions of proof in such other areas such as courts of law, everyday conversation and physics. Proofs given by physicists admit degrees.

In physics, two proofs of the same assertion have different degrees of correctness. Open any physics book, and you will verify the common occurrence of varying degrees of proofs. For example, the celebrated Peierls argument of statistical mechanics was, in a strict mathematical sense, meaningless when it was first proposed by Sir Rudolph Peierls, and remained mathematically meaningless as it underwent a number of new proofs in the sense accepted by physicists. Each successive proof was thought to be more correct than the preceding one until a "final" proof was found, the only one accepted by mathematicians.

The axiomatic method by which we write mathematics is the only one that guarantees the truth of a mathematical assertion. Nevertheless, a discussion of the axiomatic method does not tell us much about mathematical proof. Our purpose is to bring out some of the features of mathematical thinking which are concealed beneath the apparent mechanics of proof. We will argue, largely by examples, that the description of mathematical proof ordinarily given is true but unrealistic. A great many characteristics of mathematical thinking are neglected in the formal notion of proof. They have long been known but are seldom discussed.

Our ideal of realism is taken from the phenomenology of Edmund Husserl. Many years ago Husserl gave some of the rules, worth recalling here, to be followed in a realistic description.

1. A realistic description shall bring into the open concealed features. Mathematicians do not preach what they practice. They are reluctant to formally acknowledge what they do in their daily work.

2. Fringe phenomena that are normally kept in the background should assume their importance. Shop talk of mathematicians includes words like *understanding, depth, kinds of proof, degrees of clarity,* and others. A rigorous discussion of the roles of these terms should be included in the philosophy of mathematical proof.

3. Phenomenological realism demands that no excuses be given that may lead to the dismissal of any features of mathematics, labeling them as psychological, sociological or subjective.

4. All normative assumptions shall be weeded out. Too often pur-

ported descriptions of mathematical proof are hidden pleas for what the author believes a mathematical proof should be. A strictly descriptive attitude is imperative, although difficult and risky. It may lead to unpleasant discoveries: for example, to the realization that absolutely no features are shared by all mathematical proofs. Furthermore, one may be led to admit that contradictions are part of the reality of mathematics, side by side with truth.

Proof by Verification

What is a "verification?" In the ordinary sense, a "verification" of an argument establishes the truth of an assertion by listing all possible cases. Verification is one of several *kinds* of mathematical proof. A non-mathematician may believe that a proof by verification is the most convincing of all. An explicit complete list of all cases appears to be irrefutable. Mathematicians, however, are not impressed by irrefutability alone. They do not deny the validity of a proof by verification, but they are seldom satisfied with such a proof. Irrefutability is not one of the criteria by which a mathematician will pass judgment on the *value* of a proof. The value of a proof is more likely to depend on whether or not a given proof can be turned into a proof technique, that is, on whether the proof can be viewed as one instance of a type of proof, suited to proving other theorems.

Proofs by verification tend to be unsuited to extrapolation. A very simple example of a mathematical theorem for which a proof by verification could be given, but is instead avoided, is the theorem of the boxes. This theorem states that, whenever each of $n + 1$ or more marbles is to be arbitrarily placed into one of n given boxes, then at least two marbles will necessarily be placed in the same box.

This statement is so intuitive as to defy proof. Observe, however, that the evidence of this theorem is one of a particular *kind*. The theorem of the boxes is the simplest instance of an existence proof. In other words, no method is given to determine which of the n boxes will end up having at least two marbles.

To a skeptical listener (or a computer), neither willing to accept an existence proof, the intuitive argument must be replaced by a verification. A verification of the theorem of the boxes can be given, but it will be neither simple nor enlightening. It will be an algorithm that lists all ways of placing $n + 1$ marbles into n boxes, and then verifies that, for each such placement, at least one box contains at least 2 marbles. One must resort to induction to come up with an algorithm of this type, and this algorithm will be laughed at by mathematicians.

Why is the existence proof of the theorem of the boxes preferable to a verification? Because the existence proof is endowed with the light of a universal principle, unmatched by a proof by verification. A cursory look at the history of the theorem of the boxes will confirm the superiority of the existence proof.

The existence proof of the theorem of the boxes has been the starting point for the discovery of an array of deep theorems in combinatorics called theorems of Ramsey type.[1] Verification of any theorem of Ramsey type, which consists of listing all cases, is possible but only *in principle*. This verification requires a computational speed beyond the reach of our fastest computers.

Proofs of theorems of Ramsey type are existence proofs, ultimately relying on the theorem of the boxes. All these proofs are nonconstructive; nevertheless, they give incontrovertible evidence of the possibility of verification. Thus, proofs of theorems of Ramsey type are an example of a possibility that is made evident by an existence proof, even though such a possibility cannot be turned into reality.

Is All Verification a Proof?

Not all verifications are accepted as proofs despite a number of successes of proof by verification. The glamorous case of a verification that falls short of being accepted as proof — despite its undeniable correctness — is the computer verification of the four-color conjecture.

The mathematical underpinnings of such a verification were laid out long ago by the Harvard mathematician George David Birkhoff. After several failed attempts to carry out Birkhoff's program, out of desperation the aid of the computer was enlisted. Illinois mathematicians Haken and Appel set up a clever computer program which successfully carried out the lengthy sequence of verifications that Birkhoff had prescribed, and thereby established the truth of the four-color conjecture. The Haken-Appel "proof" was the first verification of a major mathematical theorem by computer.

Mathematicians have been ambivalent about such a verification. On the one hand, mathematicians profess satisfaction that the conjecture has been settled. On the other hand, the behavior of the community of mathematicians belies this feeling. Had mathematicians been truly satisfied with the computer verification of the four-color conjecture, no one would have felt the need for further verifications. But we have been witnessing an unending array of computer verifications, each quite different from the other, and each claiming to be "simpler" than the others. Every new computer program for the four-color conjecture corrects some minor oversight in the preceding one. Remarkably, these oversights are not serious enough to invalidate the verifications.

If we believed that one verification of the four-color conjecture is definitive, then later verifications would be considered a waste of time, a "machismo" game mathematicians like to play. But this is not what has happened since the work of Haken and Appel. Far from viewing successive verifications as a frivolous pastime, the mathematical community follows them with passionate interest. The latest verification invariably gets wide publicity and is subjected to intense scrutiny.

The evidence is therefore overwhelming that no computer verification of the four-color conjecture will ever be accepted as definitive. Mathematicians are on the lookout for an argument that will make all computer programs obsolete, one that will uncover the still hidden *reason* for the truth of the conjecture. It is significant that Sami Beraha, the foremost living expert on this conjecture, maintains to this day that the conjecture is "undecidable" in some sense or other despite all the computer evidence to the contrary.

The example of the four-color conjecture leads to an inescapable conclusion: not all proofs give satisfying reasons why a conjecture should be true. Verification is proof, but verification may not give the *reason*. What then do we mean by *the reason*?

Let us consider another example of verification. The classification of simple Lie groups some one hundred years ago by Cartan and Killing relies on the verification that only a finite number of configurations of vectors can be found that meet certain conditions. To this day, their verification has not been significantly improved upon, nor has the need been felt for another proof. But the Cartan-Killing classification of simple Lie groups failed to give the reason for some of the groups on the list. Cartan's list contains the Lie groups everyone expected to find, but in addition it contains five Lie groups which at the time of their discovery did not seem to conform to any pattern.

In this case too, had mathematicians been satisfied with just a list, no further work related to the classification of simple Lie groups would have been done. But this is not what happened. The existence of five "exceptional" Lie groups became a thorn in every algebraist's flesh, a rude reminder of the arbitrariness of events of the real world, an arbitrariness from which mathematics is meant to provide salvation. It became a matter of mathematical honor to find a reason for the existence of the five exceptional Lie groups.

And so, a long sequence of mathematical research papers began to appear which studied the exceptional Lie groups from every conceivable point of view. The unstated objective of these investigations was to find this reason. Eventually, MIT mathematician Bertram Kostant[2] unraveled the mystery of the exceptional Lie groups by a daring leap of faith. It turned out that the exceptional Lie groups are not the only seemingly arbitrary phenomenon in Lie theory. There is in this theory another seemingly arbitrary phenomenon: the special orthogonal group of dimension 8, or SO_8, has a very unusual property. Its covering group, or spin group, has an outer automorphism. This phenomenon happens only in dimension 8. Kostant successfully conjectured that the two phenomena must be related, and he found the *reason* for the existence of the five exceptional Lie groups in the outer automorphism of the orthogonal group in dimension 8 by a *tour de force* that remains to this day a jewel of mathematical reasoning. Once more, we are led to the conclusion that mathematicians are not satisfied with proving conjectures. They want the *reason*.

Theorems or Proofs?

Philosophers of all times have worried about problems of ontological priority. What comes first? What are the primary components of the world? Mathematicians worry about a miniature version of this problem. What is mathematics primarily made of? Roughly speaking, there are two schools.

The first school maintains that mathematics consist primarily of facts, facts like "there are only five regular solids in space." The facts of mathematics disclose useful features of the world. Never mind how these facts are obtained, as long as they are true.

The second school maintains that the theorems of mathematics are to be viewed as stepping stones, as more or less arbitrary placeholders that serve to separate one proof from the next. Proofs are what mathematics is primarily made of, and providing proofs is the business of the mathematician.

Which of these two schools shall we join? Let us consider one example that supports the second choice and one that supports the first.

No one will seriously entertain the notion that the statement of Fermat's last theorem is of any interest whatsoever. Were it not for the fact that Fermat's conjecture stood unsolved amid a great variety of other Diophantine equations which could be solved using standard techniques, no one would have paid a penny to remember what Fermat wrote in the margin of a book. What is remarkable about Fermat's last theorem is the proof. Wiles' proof appeals to an astonishing variety of distinct pieces of mathematics: to a totally unrelated conjecture made fifty years ago in the seemingly distant area of algebraic geometry, to the theory of elliptic functions, a theory that originated in the study of planetary motion, to a variety of other theories worked out over the years for unrelated purposes. The pieces of the puzzle were brought together by the magic of proof to yield the key to Fermat's secret.

The proof of Fermat's last theorem is a triumph of collaboration across frontiers and centuries. No domain of intellectual endeavor other than mathematics can claim such triumphs. But the magnitude of the triumph stands out in stark contrast to the insignificance of the statement. Insignificance of statement is a common occurrence in number theory. In number theory, the value of a theorem strictly depends on the difficulty of the proof. Perform the following thought experiment. Imagine that the theorems of number theory were suddenly to become as easy to prove as the theorems of plane geometry. Were this to happen, number theory would almost certainly find itself quickly demoted from its present exalted position to one that is indulgently granted to the theory of Latin squares. This thought experiment conclusively shows that number theory is a field consisting primarily of proofs.

Eager as we are to promulgate general conclusions, we may be tempted to conclude that the history of Fermat's last theorem is typical. Proofs come before theorems. Such a conclusion is disproved by another field of mathematics, namely, geometry.

By and large theorems of geometry state facts about the world. The information provided by theorems of geometry is valuable in unpredictable ways since geometry is rich in applications outside of mathematics. The relevance of a geometric theorem is determined by what the theorem tells us about space, and not by the eventual difficulty of the proof. The proof of Desargues' theorem of projective geometry comes as close as a proof can to the Zen ideal. It can be summarized in two words: "I see!" Nevertheless, Desargues' theorem, far from trivial despite the simplicity of its proof, has many more applications both in geometry and beyond than any theorem in number theory, maybe even more than all the theorems in analytic number theory combined.

On the basis of these two examples are we now tempted to jump to the conclusion that, in geometry, theorems are more important than proofs, whereas the opposite is true for number theory? Appearances are deceptive, and so are the above examples. To bring out the deception, let us ask ourselves whether the sharp formal distinction between theorem and proof, which seems unassailable, makes sense.

"Pretending" in Mathematics

G. H. Hardy wrote that every mathematical proof is a form of debunking. We propose to change one word in Hardy's sentence: Every mathematical proof is a form of *pretending*.

Nowhere in the sciences does one find as wide a gap as that between the written version of a mathematical result and the discourse that is required in order to understand the same result. The axiomatic method of presentation of mathematics has reached a pinnacle of fanaticism in our time. A piece of written mathematics cannot be understood and appreciated without additional strenuous effort. Clarity has been sacrificed to such shibboleths as consistency of notation, brevity of argument and the contrived linearity of inferential reasoning. Some mathematicians will go as far as to pretend that mathematics *is* the axiomatic method, neither more nor less.

This pretense of "identifying" mathematics with a style of exposition is having a corrosive effect on the way mathematics is viewed by scientists in other disciplines. The impenetrability of mathematical writing has isolated the community of mathematicians. The mistaken identification of mathematics with the axiomatic method has led to a widespread prejudice among scientists that mathematics is nothing but a pedantic grammar, suitable only for belaboring the obvious and for producing marginal counterexamples to useful facts that are by and large true.

Do not get me wrong. I am not condemning the axiomatic method. There is at present no viable alternative to axiomatic presentation if the truth of a mathematical statement is to be established beyond reasonable doubt.

One feature of the axiomatic method has been hidden from inquisitive eyes. For lack of a better language, we will call this the *exchangeability of theorem and proof.* From the point of view of formal logic, the claim that theorem and proof can sometimes be exchanged without affecting the truth or the value of a piece of mathematics seems outrageous. Everybody knows that a proof is a one-way sequence of logical steps leading to the desired conclusion. On what grounds can the preposterous claim be made that a theorem and its proof are exchangeable?

Let us go back to Fermat's last theorem. The recently completed proof requires a large number of intermediate steps. A great many of these steps are number theoretic lemmas which are given substantial proofs of their own. Some such lemmas are currently in the process of being granted the status of genuine new theorems of independent interest. Perform the following thought experiment. Imagine that Fermat had stated his conjecture in the guise of one of these intermediate lemmas. The statement might not have been quite as striking, but if the lemma were properly chosen, the proof of the lemma would be fully equivalent to the proof of the original conjecture. Any number theorist will rattle off on request an unending string of plausible sounding conjectures, each of them equivalent to Fermat's, and each of them requiring an argument fully equivalent to the argument provided by Wiles and his collaborators.

We are led to observe that the actual statement that Fermat gave of his conjecture is irrelevant to the proof. Although Wiles sets out to prove Fermat's conjecture — an event beyond dispute that has been widely reported in the media — he was in reality out to prove something else. Any one of several lemmas in Wiles' proof can be taken as a triumph of equal magnitude to the proof of Fermat's conjecture. No single part of the proof stands out as being preferable to any other, except for reasons of tradition and publicity.

But if the statement of Fermat's conjecture is irrelevant, then what is the point of all this work? The error lies in assuming that a mathematical proof, say the proof of Fermat's last theorem, has been devised for the explicit purpose of proving what it purports to prove. We repeat, appearances are deceptive. The actual value of what Wiles and his collaborators did is far greater than the proof of Fermat's conjecture. The point of the proof of Fermat's conjecture is to open up new *possibilities* for mathematics. Wiles and his collaborators show that the conjectures of Taniyama and Weil do indeed have the power that they were suspected to have. They rekindle our faith in the central role of the theory of elliptic functions in mathematics. They develop a host of new techniques that will lead to further connections betweeen number theory and algebraic geometry. Future mathematicians will discover new applications, they will solve other problems, even problems of great practical interest, by exploiting Wiles' proof and techniques. To sum up, the value of Wiles' proof lies not in what it proves, but in what it has *opened up*, in what it will make possible.

Every mathematician knows instinctively that this opening up of possibilities is the real value of the proof of Fermat's conjecture. Every mathematician knows that the computer verification of the four-color conjecture is of a considerably lesser value than Wiles' proof because the computer verification fails to open up any significant mathematical possibilities. But most mathematicians will *pretend* that the value of a proof, as well as its future possibilities, are non-mathematical terms devoid of formal meaning, and will thus be reluctant to engage in a rigorous discussion of the roles of value and possibility in a realistic description of mathematics.

A rigorous version of the notion of possibility should be added to the formal baggage of metamathematics. One cannot pretend to disregard possibility by arguing that the possibilities of a mathematical result lie *concealed* beneath formal statements. Nor can one dismiss the notion of possibility on the ground that such a notion lies beyond the reach of present day logic. The laws of logic are not sculpted in stone.

A realistic look at the development of mathematics shows that the *reasons* for a theorem are found only after digging deep and focusing on the possibilities of the theorem. The discovery of hidden reasons is the work of the mathematician. Once such reasons are found, the choice of particular formal statements to express them is secondary. Different but exchangeable formal versions of the same reason can and will be given depending on circumstances. In the search for the authentic reasons of a mathematical phenomenon, theorem and proof play the role of Tweedledum and Tweedledee. In this sense we may assert that theorem and proof are exchangeable.

Now we seem to be making a universal claim on the evidence of the one example of the proof of Fermat's last theorem. What happens to the thesis of exchangeability of theorem and proof in the example of Desargues' theorem, which is important but intuitively evident?

Let us take a closer look at Desargues' theorem. The most thorough treatment of this theorem is to be found in the first of the six volumes of Baker's *Principles of Geometry*.[4] After an argument that runs well over one hundred pages, Baker shows that beneath the statement of Desargues' theorem, another far more interesting geometric structure lies *concealed*. This structure is nowadays called the Desargues configuration. An explanation of the Desargues configuration in terms of theorems and proofs is lengthy and unsatisfactory. The Desargues configuration is better understood by meditating upon a figure displaying incident straight lines, more than 50 straight lines if I remember correctly. Once the *Ideenkreis* of the Desargues configuration is intuitively grasped, one understands the reasons that lay concealed beneath the statement of Desargues' theorem. One also realizes that for purposes of formal argument Desargues' theorem may be replaced by any one of several equivalent statements (some of them sounding quite different) which one obtains by inspecting Desargues' configuration.

The role of Desargues' theorem was not understood until the discovery of Desargues' configuration. For example, the fundamental role of Desargues' theorem in the coordinatization of synthetic projective geometry can only be understood in the light of the Desargues configuration. Thus, a formal statement as simple as Desargues' theorem is not quite what it purports to be. While it pretends to be definitive, in reality it is only the tip of an iceberg of connections with other facts of mathematics. The *value* of Desargues' theorem and the *reason* why the statement of this theorem has survived through the centuries, while other equally striking geometrical theorems have been forgotten, is in the realization that Desargues' theorem opened a *horizon of possibilities* that relate geometry and algebra in unexpected ways. In conclusion: what an axiomatic presentation of a piece of mathematics *conceals* is at least as relevant to the understanding of mathematics as what an axiomatic presentation *pretends* to state.

Are there Definitive Proofs?

It is an article of faith among mathematicians that after a new theorem is discovered, other simpler proofs of it will be given until a definitive one is found. A cursory inspection of the history of mathematics seems to confirm the mathematician's faith. The first proof of a great many theorems is needlessly complicated. "Nobody blames a mathematician if the first proof of a new theorem is clumsy," said Paul Erdös. It takes a long time, from a few decades to centuries, before the facts that are hidden in the first proof are *understood*, as mathematicians informally say. This gradual bringing out of the significance of a new discovery takes the appearance of a succession of proofs, each one simpler than the preceding. New and simpler versions of a theorem will stop appearing when the facts are finally understood. Unfortunately, mathematicians are baffled by the word "understanding," which they mistakenly consider to have a psychological rather than a logical meaning. They would rather fall back on familiar logical grounds. They will claim that the search for *reasons*, for an *understanding* of the facts of mathematics can be explained by the notion of simplicity. Simplicity, preferably in the mode of triviality, is substituted for understanding. But is simplicity characteristic of mathematical understanding? What really happens to mathematical discoveries that are reworked over the years?

Let us consider two examples.

Example one. The first proof of the pointwise ergodic theorem, which the older George David Birkhoff of Harvard found in answer to a challenge of the young John von Neumann of Princeton,[5] was several pages long. The proof of the same theorem, indeed the proof of a much more general version, given by Adriano Garsia in 1964[6] took up half a page, including all details. Garsia's successful insight is a prime example of simplification almost to the point of triviality.

Example two. Sometime in the fifties, Hans Lewy of Berkeley discovered the first example of a partial differential equation having no solutions whatsoever.[7] In the succeeding thirty years, the idea hidden underneath Lewy's example was gradually made explicit until the *reason* for such an impossibility became clear. Nonetheless, the fact that a partial differential equation may have no solutions remains to this day non-trivial, despite the fact that the reasons for such an occurrence are now understood.

In each of these examples the process of simplification required the hard work of generations of mathematicians. Shall we then assert that mathematicians spend their lives in search of simplification?

Garsia's proof of the pointwise ergodic theorem appears to be trivial, but only *a posteriori*. It is hard to imagine anyone discovering such a proof without foreknowledge of the history of the theorem. On the other hand, Hans Lewy's mysterious discovery was ensconced into a new theory of commutators of differential operators, which eventually disclosed the hidden *point* of Hans Lewy's example. The theory is elegant and definitive, but far from simple.

There are other embarrassing counterexamples to our faith in simplicity. Perhaps the oldest and the most dramatic such counterexample is the fundamental theorem of geometry, going back to von Staudt in the early nineteenth century, stating the equivalence of synthetic and analytic projective geometry.[8] No significant progress has been made in simplifying von Staudt's proof. Even today, a full proof of von Staudt's theorem takes no less than twenty pages, including a number of unspeakably dull lemmas. Every geometer is dimly aware of the equivalence of synthetic and analytic projective geometry; however, few geometers have ever bothered to look up the proof, let alone remember it. Garrett Birkhoff, in his treatise on lattice theory,[9] a book purporting to deal precisely with this and related topics, gives the statement of von Staudt's theorem, and then gingerly refers the reader to a proof by Emil Artin that was privately distributed in mimeographed form in the thirties at the University of Notre Dame.[10]

Von Staudt's theorem was so far removed from the mathematical mainstream that in the thirties von Neumann rediscovered it from scratch, giving much the same proof as von Staudt's while developing his theory of continuous geometries (I have been told by Stan Ulam that von Neumann, upon learning of von Staudt's work almost a century before, fell into a fit of depression). Philosophers of mathematics have speculated that the difficulty in simplifying von Staudt's proof may be due to the fact that the assertion is only valid in dimension three or greater, while in dimension two a plethora of awkward counterexamples has been found, the nasty non-Desarguian projective planes. From time to time some courageous mathematician will exhume von Staudt's theorem and renew the attempt to find a transparent proof. The latest is due to Mark Haiman of San Diego, who succeeded in giving a brilliant and short conceptual proof at the cost of making one tiny simplifying assumption that still requires a lengthy proof.

We would like to believe that such instances of theorems whose proofs resist simplification are rare. Luckily, no theorem discovered before 1800 belongs in the same class as von Staudt's theorem. Most mathematics discovered before 1800, with the possible exception of some very few statements in number theory, can nowadays be presented in undergraduate courses, and it is not too far-fetched to label such mathematics as simple to the point of triviality.

On the other hand, in the twentieth century we observe the uncomfortably frequent occurrence of easily stated results whose proofs run over hundreds of pages. An example is the classification of finite simple groups. The variety of intermediate theorems needed for this classification is so large as to defy the mental powers of any individual mathematician. Nonetheless, the classification of finite simple groups is no mere brute force verification. The arguments leading to drawing the complete list, long and inaccessible as they are, have the merit of "explaining" in a conceptually satisfactory way the *reason* why the only existing finite simple groups are what they are. The overall argument is convincing, even if impossible to follow, and it is conceivable that no further simplifications will be forthcoming.

The example of Desargues' theorem is also disquieting. Here, the opposite of a simplification occurred as the centuries went by. Later work inspired by Desargues' theorem made it look more complex than it first appeared. In Fermat's last theorem we note yet another phenomenon. The search that mathematicians are now beginning to perform into the proof of Fermat's last theorem is meant to discover "what" it is that is "really" being proved. It is a search that will keep mathematicians busy for a long time.

The Secret Life of Mathematics

Hardy was right after all: mathematicians are out to debunk the fakery that lies concealed underneath every logically correct proof. But they will not admit that their task is one of debunking; they will rather pretend that they are busy proving new theorems and stating new conjectures in compliance with the canons of present day logic.

Every theorem is a complex of hidden possibilities. In the example of Desargues' theorem some of these possibilities were eventually brought out by the discovery of the Desargues configuration. Behind Hans Lewy's example of a differential equation without solutions lurked untold facts about differential operators. The pointwise ergodic theorem called for understanding the full range of possibilities that had been opened by Lebesgue's ideas, an understanding that was effectively exploited in Garsia's proof. And the proof of Fermat's last theorem foreshadows an enormous wealth of mathematical possibilities.

It does not seem unreasonable to conclude that the notion of *possibility* will be fundamental in future rigorous discussions of the nature of mathematics. Disquieting as this prospect may appear, there are other, equally fundamental and equally disquieting notions that call for rigorous treatment, and that present day logicians have yet to admit into their fold. For example, mathematical proofs come in different *kinds* that need to be classified. The notion of *understanding*, that is used in informal discussion but quashed in formal presentation, will have to be given a place in the sun; worse still, our logic will have to be modified to accommodate *degrees* of understanding. Lastly, the notion of *evidence* will have to be given a formal standing that puts it ahead of the traditional notion of truth. The characteristic features of mathematical evidence, many of them refractory to treatment in formal language, will have to be accurately described. Some additional features might be the notions of *value*, of *reasons*, and most of all, the phenomenon of *concealment* of a theorem by another theorem that occurs everywhere in mathematics. Will anyone ever undertake this daunting task?

CHAPTER XII

Syntax, Semantics, and the Problem of the Identity of Mathematical Items[1]

The items of mathematics, such as the real line, the triangle, sets, and the natural numbers, share the property of retaining their identity while receiving axiomatic presentations which may vary radically. Mathematicians have axiomatized the real line as a one-dimensional continuum, as a complete Archimedean ordered field, as a real closed field, or as a system ofbinary decimals on which arithmetical operations are performed in a certain way. Each of these axiomatizations is tacitly understood by mathematicians as an axiomatization of the *same* real line. That is, the mathematical item thereby axiomatized is presumed to be the *same* in each case, and such an identity is not questioned. We wish to analyze the conditions that make it possible to refer to the *same* mathematical item through a variety of axiomatic presentations.

The analysis of sameness of mathematical items, as we shall call the constructs of mathematics, points to a closely related problem, one that was first recognized in symbolic logic but that has a far wider scope. A presentation of a mathematical system leading up to the definition of an item is of necessity syntactical, that is, it is given by axioms and rules of inference. The axioms and rules of inference are intended to characterize a class of mathematical items consisting of sets with some additional structure (such as groups, manifolds, etc.). Any structure that satisfies the axioms is said to be a *model* for the axioms, and the description of all models is the semantic interpretation of the theory. The problem of identity in mathematics can be viewed as the problem of explaining how disparate syntaxes can have the same semantics, that is, the same models.

This duality of syntactic and semantic presentation is shared by all mathematical theories.

We survey some examples of semantic and syntactic description of mathematical theories, where one may have syntax without semantics, or semantics without syntax. We illustrate by examples the variety of axiomatic presentations that can be given to the same mathematical item.

An adequate understanding of mathematical identity requires a missing theory that will account for the relationships between formal systems that describe the same items. At present, such relationships can at best be heuristically described in terms that invoke some notion of an "intelligent user standing outside the system."

We stress that the problems of referential identity and of syntax/ semantics, though born out of mathematics, are universal to all of science.

Syntax and Semantics

Syntax and semantics in a formal system are best understood by the one example that gave rise to these notions; that is, the propositional calculus or Boolean algebra or the theory of distributive lattices, as it has been variously called, and its extension to the predicate calculus by the addition of quantifiers.

The algebra of sets, and the fact that sets can be manipulated by algebraic operations, fascinated mathematicians and philosophers since Leibniz. In the nineteenth century, in the work of Boole, Schröder, Peirce, Peano, and others, we find the first attempts at axiomatization.

The ultimate distillation of these ideas was the notion of a lattice satisfying the distributive law; that is, of a syntactic structure consisting of two operations called join and meet (in symbols, respectively, $\vee$ and $\wedge$), satisfying the usual idempotent, commutative, associative, and absorption laws, and in addition satisfying the distributive law (identity): $x \vee (y \wedge z) = (x \vee y) \wedge (x \vee z)$.

It is a great achievement of mathematics to have shown that the distributive law alone is the syntax for the structure consisting of a family of sets closed under unions and intersections. What is more, a model can be constructed for every distributive lattice using the syntactical description. The theory of sets is made into an algebraic system (in the sense of universal algebra) subject only to identities, and all models of such an algebraic system are shown to be families of sets. One cannot hope for a more definitive way of getting to the semantics.

This result, first clearly stated by Garrett Birkhoff and Marshall Harvey Stone, is the paradigm of formalization of a mathematical theory.[2] Elementary though it has become after successive presentations and simplifications, the theory of distributive lattices is the ideal instance of a mathematical theory, where a syntax is specified together with a complete description of all models, and what is more, a table of semantic concepts and syntactic concepts is given, together with a translation algorithm between the two kinds of concepts. Such an algorithm is a "completeness theorem."

On the blueprint of the propositional calculus an as yet unformalized program for a syntax/semantic analysis has been stated that in principle should be applicable to any mathematical theory. This program consists broadly of the following:

1. A criterion of syntactical equivalence ("cryptomorphism") applicable to all axiomatic presentations. At our present state of understanding of formal systems, it is not clear what such a criterion could be, even though in mathematical practice there is never any doubt as to the equivalence of different presentations of the same theory.

2. A "classification" of the semantic models of the theory independent of the choice of a particular axiomatization.

3. A "completeness theorem" relating truth in all models ("semantic truth") to truth as proof from axioms ("syntactic truth").

We begin by listing a few examples of active mathematical theories in which either the syntax or the semantics is missing.

The syntactical development of the predicate calculus by the Hilbert ϵ-operator has resisted so far all attempts at semantic interpretation. Briefly, the Hilbert ϵ-operator is a successful syntactic formalization of the notion "the individual variable x such that $F(x)$ is true, if any," where $F(x)$ is a predicate. One replaces the sentence $(\exists x)\ F(x)$ by the proposition $F(\epsilon(F(x)))$. [3] Likewise, before the work of Kripke and others, the semantics of intuitionistic and modal logics were not known.

These are examples of syntactic structures that have no semantics. The discovery of all models of intuitionistic logic led to a major advance in logic, namely, the realization that intuitionistic logic is a syntactical presentation of Paul Cohen's idea of forcing.[4]

Since von Neumann it has been agreed that closed subspaces of a Hilbert space are the quantum mechanical analogs of probabilistic events. The theory of closed subspaces of Hilbert space is an elementary semantics of quantum mechanics. However, numerous attempts at developing a "logic of quantum mechanics" have failed. No one has yet been able to develop a syntactical presentation of the events of quantum mechanics. The theory of modular lattices, or the theory of orthomodular lattices, have so far met with little success because their semantics does not agree with the practice of quantum mechanics.

Throughout mathematics, structures that are semantically described occur more frequently than structures that are syntactically described. This is not surprising: the syntax of a structure is often an afterthought.

The case of mathematical logic, where the syntax was developed before the semantics, is exceptional because of the unusual history of the subject. An instance of a semantics without syntax is the theory of multisets. A multiset of a set S is a generalization of a subset of S, where elements are allowed to occur with multiplicities. Multisets can be added and multiplied; however, a characterization by algebraic operations of the family of multisets of a set S — an analog of what Boolean algebra is for sets — is not known at present. This problem is of more than academic interest. There is a deep duality between the algebra of sets and the algebra of multisets that a syntactical description may well elucidate.

Another example of a semantic theory where a syntax is missing is probability theory. Statisticians and probabilists employ informal syntactical presentations in thinking and speaking to each other. When a probabilist thinks of a Markov chain, he seldom appeals to the a path space in which a Markov chain may be realized. The statistician who computes with confidence intervals and significance levels seldom appeals to measure-theoretic models. In a syntactical presentation of probability, joint probability distributions are similar to truth values in the predicate calculus. Kolmogorov's consistency theorem is a completeness theorem relating the syntax and semantics of probability.

The Variety of Axiomatic Presentations of an Item

There need not be a one-to-one correspondence between axiomatic systems (syntax) and models (semantics). An axiomatic system intended for a specific model may turn out to have other models, sometimes called non-standard models. Conversely, the same mathematical theory may be presented by a variety of cryptomorphic axiom systems.

It is impossible to spell out all the axiom systems of a theory. The theory of groups may be axiomatized in innumerable ways, some of which may even use ternary operations. Every such axiomatization is guided by a previous understanding of the notion of group. The concept of group is learned from a specific axiom system. A student will use axiomatics as a crutch to be forgotten as soon as he gains familiarity with the theory and is thereby freed from depending on any particular axiomatic formulation. We will call the understanding of the notion of group which is free from a choice of an axiom system a *pre-axiomatic grasp*. A pre-axiomatic grasp of the notion of group cannot be gleaned from familiarity with specific groups and their "common" properties.

The fact that a variety of axiom systems exists for the theory of groups does not privilege of any one system, but on the contrary presupposes a pre-axiomatic grasp of the notion of group.

The understanding of a mathematical theory is not the result of enlightened familiarity with an axiom system given once and for all. To the mathematician, an axiom system is a window through which an item, be it a group, a topological space or the real line, can be viewed from a different angles that will reveal heretofore unsuspected possibilities.

The real line has been axiomatized in at least six different ways, each one appealing to different areas of mathematics, to algebra, number theory, or topology. Mathematicians are still discovering new axiomatizations of the real line. Such a wealth of axiomatizations shows that to the mathematician there is only *one* real line. The more extravagant the axiomatizations become, the firmer is the mathematician's preaxiomatic grasp of the one and only real line. The need for further axiomatizations is motivated by the discovery of further properties of the real line. The mathematician wants to find out what *else* the real line can be. He wants ever more perspectives on one and the same real line. Such new perspectives are precisely the various axiomatizations, new syntactic systems always having the same model. The need for more perspectives on one and the same mathematical item is is part of the formal structure of mathematics.

Every axiomatic system for the line has a pretension of definitiveness that is belied by the openendedness of the properties of the real line. This openendedness is a property of all mathematical items.

Conclusion

A full description of the logical structure of a mathematical item lies beyond the reach of the axiomatic method as it is understood today.

Two or more axiom systems for the real line can be recognized to have the *same* real line as their model. This recognition is not carried out within any axiom system or by comparing different axiom systems; it is the preaxiomatic grasp of the real line. The notion of a preaxiomatic grasp has the following consequences:

1. The real line, or any mathematical item, is not fully given by any one specific axiom system.

2. The totality of possible axiom systems for the real line cannot be foreseen. Any mathematical item allows an open-ended sequence of presentations by new axiom systems. Each such system is meant to reveal new features of the mathematical item.

3. Learning about the real line is not a game played with axioms whereby skill is developed in drawing consequences. Rather, the choice of properties to be proved and the organization of the theory is guided by a pre-axiomatic grasp. Without a preaxiomatic grasp, no axiomatic theory can make *sense*.

4. Even though a concept may be learned by working through one axiomatic approach, that particular axiomatic approach will be shed after familiarity is gained. In learning by a particular axiomatic system, a concept is revealed whose understanding lies *beyond* the reach of the axiom system by which it has been learned.

CHAPTER XIII

The Barber of Seville, or The Useless Precaution[1]

Dear Professor Spalt,

Thank you very much for your comments on our article concerning the identity of mathematical items. We should like to outline a possible response to your arguments.

We fully agree with you that mathematical items have none of the substantiality of everyday items, and we agree one hundred percent with your statement that mathematical items are no mere substances, they are mere *relata*. It has never crossed our minds to assign "substantiality" to mathematical items. We also agree with your statement that the mathematical items of today are given in relational form as axiom systems. But one must distinguish between the way of givenness of the items of mathematics and the identity of mathematical items. It is true, as you say, that mathematical concepts are defined by relational systems. But it would be an error to identify the items with the relational systems that are used to define them. I can define the triangle in many ways; however, no definition of the triangle is *the same* as the "item" triangle. There are many ways of defining the real line, but all these definitions define something *else*, something that is nevertheless distinct from the relations that are used in order to define it, and which is endowed with an identity of its own.

We disagree with your statement that the ideal form of knowledge in theoretical mathematics is the theorem (and its proof). What we believe to be true is that the theorem (and its proof) is the ideal form of *presentation* of mathematics. It is, in our opinion, incorrect to identify the manner of presentation of mathematics with mathematics itself. We would be presumptuous to believe that the axiomatic system, of which we are so proud, is the definitive way of presenting mathematical knowledge.

Your rhetorical question, "What does a pre-axiomatic grasp grasp?" can be answered by a historical example. Cauchy had no rigorous definition of the real line or the integral. Nonetheless, what he wrote about the real line and the integral is still considered valid and has been adopted in the rigorous axiom systems for the real line and the integral later devised by Dedekind and Riemann. How is it possible to incorporate Cauchy's results on the real line into Dedekind's axiom system without having a pre-axiomatic grasp of the fact that both mathematicians were dealing with the *same* real line?

Wiener axiomatized the group law by taking xy^{-1} as the basic operation, and his axiomatization is quite different from any of the other axiom systems for groups. Nonetheless, "it has never fooled anybody" into believing that what Wiener was talking about was anything but a group. How do we know that the axiomatizations of the real line by Dedekind cuts and as a complete Archimedean ordered field describe the same real line? The answer given by some people to this question, namely, that we map each axiom system into the other, is not honest. We perform this mapping only after we have realized that the two axiom systems describe the same notion.

This brings us to the crux of our disagreements, which is your use of the word "psychological." You write, ". . . this only grasps the psychological topic of motivation (to take traditional pedagogical language) or of desire (to take some modern philosophical language)." And later: "To sum up. . . Rota/Sokolowski/Sharp illicitly mix up a psychological and a mathematical topic. . . ."

We think your use of the word "psychological" in these contexts is incorrect. What can this word possibly mean? Psychologists have never dealt with these matters, and they never will. Could it mean that the matters we deal with are dependent on individual whim? Clearly not, since otherwise, it would be impossible to discuss them. Could it mean that the problems we are dealing with are "mental?" But all mathematics is mental.

What you call psychological we call philosophical. Motivation and desire are essential components of mathematical reasoning. We have no right to dismiss these aspects of mathematics as "purely psychological," any more than we would dismiss Cauchy's results on integration as "purely heuristic," because he failed to provide a full axiomatization. Dismissing as psychological anything that is not explainable in terms of axiomatics is a way of disposing of uncomfortable problems. The axiomatic system, despite its advantages over other methods for the presentation of mathematics, has a built-in wishful thinking, as if an axiom system were to bring into being the notions that are being defined. Mathematicians proceed as if the notions of mathematics sprang *ursprünglich* out of the axioms. Some philosophers of mathematics, even the greatest, have dismissed under the label of "psychological" anything that threatens this make-believe.

You write: "It is wrong (today) to claim that there are mathematical items beyond any axiom system." We can neither agree nor disagree with this statement. We agree that there is no way of dealing with mathematical items rigorously except through axiom systems. But this is like saying that there is no way of communicating ideas except through words. Although any idea has to be expressed in sentences, the same idea may be expressed by completely different sentences. An idea is "independent" ("independent" is a dangerous word, but we find no better one) of the words that are used to express the idea. When we assert that a mathematical item is "independent" of any particular axiom system, we mean this "independence" in much the same way as the independence of ideas from language.

It is tempting at this point to draw the conclusion that we assume that mathematical items "exist" independently of axiom systems. This is not what we mean. You write: "Everybody who starts talking about mathematical items 'freed' from any axiomatic system must explain what is meant by existence." You are wrong. In discussing the properties of "mathematical items," we are in no way required to take a position as to the "existence" of mathematical items. Identity does not presuppose existence.

You are echoing a widespread prejudice. The discussion of existence of mathematical items has gone on for a long time. Some respectable thinkers have held that mathematical items exist (in some sense or other); other equally respectable thinkers have argued that mathematical items do not exist. The end result is that the opinion has made headway that it does not matter whether mathematical items exist, and probably it makes little sense to ask the question.

Perform the following *Gedankenexperiment*: imagine that someone proved beyond any reasonable doubt that mathematical items do not exist (but are nevertheless consistent). Do you think such a proof would affect the truth of any mathematical statement? Certainly not. Your statement, "To say that there is the same item axiomatized (as Rota/Sharp/Sokolowski do) inevitably presupposes that there exist some items for which it is possible to propose they are identical" is wrong. One can spend a lifetime working on mathematics without ever having any idea whether mathematical items exist, nor does one have to care about such a question. The existence of mathematical items is a chapter in the philosophy of mathematics that is devoid of consequence. We do not object to anyone who chooses to worry about existence. *Jedem Tierchen, sein Pläsirchen.* Discussions of "existence" are motivated by deep-seated emotional cravings for permanence which are of psychiatric rather than philosophical interest. Verifying the existence of anything is like asking for an authorization to proceed while failing to realize that no such authorization is required. The comedy of mathematical existence might be given the title of the well-known opera, *The Barber of Seville, or the Useless Precaution.*

CHAPTER XIV

Kant and Husserl

Kant, Husserl, and Mathematics

Few distinctions have had as much influence on the later development of philosophy than Kant's distinction between analytic, synthetic, and synthetic *a priori* statements. Few arguments in Western philosophy have been as instrumental in redirecting philosophical investigation as Kant's argument leading to his thesis of possibility of synthesis *a priori*.

Our purpose is to reexamine the standing of Kant's synthesis *a priori*. What is synthesis *a priori* about? We will attempt an answer from the vantage point of contemporary phenomenology.

Following Kantian wishes, our examples are drawn from mathematics. Mathematics has enjoyed an exemplary status among intellectual disciplines. Like music and art, it has played in our culture the role of Cassandra. Whatever will happen in the world of mind happens first in the small world of mathematics.

It is therefore no surprise to find that the problem of the "ontological standing" of the statements of mathematics is intertwined with the problem of synthesis *a priori*.

The variety of positions that have been taken on mathematics has bequeathed to us a panoply of terms, each one with a slant of its own. It may be prudent to avoid such loaded expressions; accordingly, we will adopt the neutral word "item" to denote both the assertions of mathematics and the objects of mathematics. Our concern will not be to discover what mathematics is made of; rather, our objective will be to discover the conditions of possibility of that eidetic domain that is called mathematics. The variety of terms used in the past to denote mathematical items makes a stark contrast to the dearth of examples. The philosophy of mathematics is beset with insistent repetitions of a few crude examples taken from arithmetic and from elementary geometry, in total disregard of the philosophical issues that are faced by mathematicians at work.

The shunning of relevant examples was criticized by Husserl when he laid down his rules for a phenomenological description. Husserl's injunction was to give up any *parti pris* on the nature of mathematics and to begin instead by making a pedestrian list of mathematical items together with their phenomenological characteristics, all the while resisting any attempt to conceal heterogeneity or irreconcilable features among the items in our list, and most of all, guarding against normative assertions. Contradictions shall not be be resolved by selecting a biased sample of examples but by letting the examples themselves display their common conditions of possibility.

The Kantian classification into synthetic, analytic, and synthetic *a priori* items applies to mathematics, but only with a vengeance. While some of the items of mathematics appear to be analytic, others appear instead to be synthetic or *a posteriori*. We will make no attempt to prescribe one of these alternatives for all mathematical items.

An Example from Mathematics: The Classification of Finite Simple Groups

One of the great achievements of mathematics in this century is the classification of finite simple groups, which was completed some fifteen years ago after one hundred years of research by the joint efforts of mathematicians belonging to several countries and several generations. The work of these mathematicians adds up to several thousand pages of closely packed reasoning. Thanks to their contributions, we are now able to draw a finite list that includes all simple groups that do and can exist, that is, a list of all types of symmetry. The list is short and reasonable; it contains those groups which everyone thought to be simple, plus some dozen-odd at first rather grotesque-looking groups which were discovered with enormous effort. On second glance, even these monstrously large groups that were added at the last stage of the classification reveal themselves to be put together in amusing and clever ways. The Lord may be playful, but never arbitrary.

The classification of finite simple groups is one of the great achievements of the human mind. The detailed proof that the list is complete appeals to every previously known result in the theory of groups, as well as to a variety of other mathematical techniques. Nonetheless, when viewed from the strict viewpoint of Fregean logic, the entire proof can be reduced to the definition of a group — a definition that any child can understand — together with a string of arguments starting from such a definition and following the rules of logic.

In a strict logical sense, the list of finite simple groups is an analytic statement which trivially follows from the definition of a group. The list of finite simple groups is somehow to be found "in" the definition of a group. Anyone who understands such a definition "must" (again, the use of the verb "must" is as problematic as that of "in") be able to arrive at the classification of simple finite groups by logical reasoning. And yet in our heart we know that such an assertion is misleading.

If we follow Husserl's suggestion to give a non-normative description, then we are struck by the irrefutable fact that those mathematicians who worked on the classification of finite groups did not try out one reasoning process after another on the definition of group. Some of them were poor logicians, and their purported proofs had to be corrected. Isn't this situation outrageous? Isn't it preposterous that mathematicians trying to prove a logical statement proceed in illogical ways?

What makes this observation preposterous is our prejudice as to how mathematicians work.

Guesswork in Mathematics and Science

Some of the deepest mathematical results are arrived at by guesswork. We use the term "guesswork" to denote that extraordinary happening whereby some new item of mathematics appears in the world. The term "guesswork" is a copout word. What happens is that a mathematician, or a group of mathematicians (often working at different times in history), becomes certain of the truth of a mathematical assertion long before he is able to give any argument. Guesswork is the rule, not the exception, in mathematics.

If mathematics were an analytic discipline, then the occurrence of guesswork would have to be justified by a list of other copout expressions, such as good luck, instinct, or talent.

The hidden assumption that leads us to dismiss an essential part of the mathematician's activity by the copout word "guesswork" is that the work of a mathematician should be one of two kinds: there is mathematical argument, which is regarded as belonging to logic, and there is mathematical discovery, which is regarded as belonging to psychology. By this artificial classification, one quietly disposes of guesswork by labeling it as "psychological."

Again following Husserl, we may ask: what entitles us to play favorites among equally relevant features of a mathematician's work? What is the justification, if any, for labeling some features of mathematical work as "psychological" with the purpose of dismissing them without a hearing?

The imbalance between "logical" and "psychological" features of mathematical work becomes glaring when compared to "guesswork" in physics, chemistry, or biology. Here philosophers take a different view of guesswork. They admire the guesswork of the physicist who wrests a new law from the jungle of raw nature, and they ennoble such guesswork by the title "inductive logic," a presumed and presumptuous new logic that has been waiting just around the corner for the last hundred years.

Here again we find an unjustified subdivision of the scientific enterprise into two areas. On one side we find the sciences of nature; on the other we find the sciences of mind, paramount among them mathematics. The first are thought to be heroically wrestling synthetic laws out of nature; the second are given subordinate status by being labeled "analytic."

A mathematician's guesswork is no different from guesswork in any other science. Finite simple groups were discovered by guesswork. None of them was discovered by logical reasoning starting with the definition of a group. Some of them were discovered by analyzing symmetries in natural phenomena; others were constructed by the trial and error method of all science. The exploratory work of a mathematician is indistinguishable from the exploratory work of a physicist. Both are exercises of the imagination, and in both cases guesswork of the wildest kind is required in order to produce a large enough variety of hypotheses from which one will be left standing after the acid test of reality.

Is Mathematics Analytic or Synthetic?

It has been known since Plato that in Euclidean space of three dimensions only five regular solids can exist: the tetrahedron, the cube, the octahedron, the dodecahedron, and the icosahedron.

Sometime in the nineteenth century mathematicians discovered that in a space of four dimensions six regular solids exist, and that in spaces of all dimensions higher than four, only three regular solids exist, the analogs of the cube, the tetrahedron, and the octahedron. Are these discoveries synthetic statements of fact to be confirmed by observation of nature, or are they analytic statements following by logical arguments from the axioms defining Euclidean space? The only honest answer is: they are both.

The classification of regular solids finds confirmations in chemistry and allows chemists to prescribe the behavior of crystals and other molecules. There are situations in physics in which the fact that only five regular solids exist is viewed on a par with a physical law. One may infer the existence and uniqueness of the five Platonic solids from empirical observation rather than from mathematical axioms. Even in botany one finds empirical classifications that might lead us to the regular solids.

The troubling conclusion that emerges from this example is that there is no clear boundary between "abstract" mathematical items and the empirical facts of science. A mathematical item may be discovered as a purely empirical fact. The classification of mathematical items as "ideal objects" while reserving to physical items the status of real objects is questionable. Only a prejudice would lead us to classify the straight line as an "ideal" object while the free fall of bodies is awarded the status of "real." Both these items share the same feature of being in the world, of being transcendental, as Kant was first to say.

The classification of items into real and ideal is useful, provided we stress the proviso that the same item may be sometimes real and sometimes ideal, sometimes mathematical and sometimes chemical or botanical. The term "sometimes" is indispensable in realistic description. Ambiguity is a feature of reality, and it is no wonder that we find it when considering the status of the items of mathematics.

The demarcation line between mathematics and physics is blurred. Some mathematical theorems were first taken to be laws of physics; the most famous is the central limit theorem of probability theory, of which the French mathematician Henri Poincaré stated, "Physicists believe it to be a mathematical theorem, and mathematicians believe it to be a law of physics." The central limit theorem is *both* a law of nature and a mathematical theorem. It is an elegant consequence of the definition of a stable distribution of finite variance *and* an experimental fact which any child can observe at the science museum. It helps little to ignore half the truth.

The central limit theorem is now a mathematical theorem and it was once a law of nature. Other examples of such mutations are frequent. There are laws of physics which were eventually turned into mathematical theorems. The most famous such instance is classical mechanics, a subject which was once the core of engineering at its most practical and which is now taught in mathematics departments, on a par with geometry.

We may now draw three conclusions.

1. Mathematical items can be viewed either as analytic statements derived within an axiomatic system or as facts about the natural world, on a par with the facts of any other science. Both claims are equally valid.

2. The contextual standing of an item as analytic or synthetic is not fixed. A law of physics may come to be viewed as a theorem of mathematics. A theorem of mathematics may acquire applications that may turn it into a fact of ordinary life.

3. A mathematical item is either *a priori* or synthetic depending on the context. Such a contextual ambiguity does not entitle us to infer that an item is "synthetic *a priori*."

Evidence and $\alpha\lambda\eta\theta\varepsilon\iota\alpha$

A *Leitfaden* out of the disquieting conclusion that an item may be both analytic and synthetic is found in Husserl's thesis of the *priority of evidence over truth*. Evidence is the primary logical concept. Evidence is the condition of possibility of truth. Truth is a derived notion.

Phenomenological evidence is *a logical concept*, on a par with definition or inference. The de-psychologization of evidence is a great achievement of phenomenology.

Primordial evidence, $\alpha\lambda\eta\theta\varepsilon\iota\alpha$ to the Greeks, is constitutive of mathematical reasoning. The experience that most mathematicians will corroborate, that a statement "must" be true, is not psychological. In point of fact, it is not an experience. It is the condition of possibility of our experiencing the truth of mathematics.

An experience that is shared by most mathematicians betokens the standing of evidence. After reaching evidence we go back to our train of thought and try to pin down the moment when evidence was arrived at. We pretend to locate the moment we arrived at evidence in our conscious life. But such a search for a temporal location is wishful thinking.

In light of the primacy of evidence, the assertion that much of the mathematician's work is "guesswork" becomes downright obvious.

Truth and the Axiomatic Method

What becomes of truth in a logic that takes evidence as its primary concept? And what is the standing of Tarskian truth in the light of the primacy of evidence?

By reading a proof, a mathematician can reconstruct the original evidence of a mathematical item as it was meant by whoever wrote the proof. It is a mistake to view mathematical proof as the recipe for carrying out some sort of reasoning. The purpose of proof is to make it possible for one mathematician to bring back to life the same evidence that had been previously reached by another mathematician. While proof is only one of many possible ways of resurrecting evidence, to this day proof has shown itself to be the most reliable.

Truth, in the ordinary sense of "truth of a sequence of inferences," is the recognition of the identity of my evidence with someone else's evidence. In mathematics this realization is arrived at by verifying the correctness of the successive steps in a proof. Again, truth is subordinated to evidence.

A striking inversion of priorities occurs in phenomenology. Evidence is the primary logical notion, and truth is a rhetorical device having to do with the *communication* of evidence.

Tarskian truth has to do with the recording and resurrecting of evidence; it has little to do with *adaequatio rei et intellectus*, or any of the other trite concepts of truth.

It is a wonder of wonders that mathematics has achieved perfection in the recording of evidence by means of the axiomatic method, and that mathematics is the only eidetic domain for which communication across frontiers of time and cultures has been successful.

There is no such thing as an analytic statement. Those statements that we used to label "analytic" are steps in some proof, building blocks of a contraption designed to produce evidence. The truth of such a statement is made possible by a previous realization of the evidence of the assertion. The truth of a statement in the sense of mathematical logic is a grammatical matter, having to do with a successful technique of exposition; it is unfortunate that philosophers should have been misled into believing that the tidy and efficient technique that is called mathematical logic has any bearing on $\alpha\lambda\eta\theta\varepsilon\iota\alpha$.

The dichotomythat makes it appear that an item of mathematics sometimes is an analytic deduction from a set of axioms and sometimes is a fact among facts in the world is spurious. We have instead two *modes* of evidence of the *same* item. On the one hand, the evidence of axioms and proof, and on the other hand, the less reliable evidence that a mathematical item is a fact of the world and thus to a varying extent experimentally verifiable. There is no reason why these two modes of evidence should be in conflict, since we live in one world.

Evidence, Fulfillment, and Synthesis A Priori

The thesis that evidence is transcendental is fraught with the danger that haunts all phenomenological theses. A subtle form of reductionism may be at work when "evidence" is regarded as an instantaneous process reminiscent of a light bulb being lit. A double error is at work here. First, evidence is reduced to a process in space and time. As a consequence of this reduction, a search is undertaken for physical causes of evidence, which are sought in mental processes and brain functions. Second, the primacy of evidence is taken as an excuse for avoiding details in the description of evidence.

Husserl's phenomenology of intention and fulfillment was motivated by his purpose of providing a phenomenological underpinning to the primacy of evidence, cleansed of psychologistic interpretations.

Husserl's description brings out the complexity of the phenomenon. The role of *iteration* in the fulfillment of evidence is singled out for special description. The *gradual* discovery of an item beckoning for evidence fragments into a temporality of its own. The imperfection, the lack of certainty, the insecurity of evidence are described by Husserl as features of all evidence whatsoever.

Husserl's description is confirmed by observing mathematicians at work. Mathematicians are frequently frustrated by the experience of verifying the correctness of every step of a proof without being able to attain the evidence of the statement being proved. Such frustration is the rule rather than the exception. A proof that shines with the light of evidence comes long after discovery. Thus, whereas mathematicians may claim to be after truth, their work belies this claim. Their concern is not truth, but evidence. The gradualness of mathematical evidence, together with the implicit admission of the priority of evidence that is betokened by the search for evidence, confirm Husserl's description of evidence in his phenomenology of intention and fulfillment, as well as his thesis of the primacy of evidence over truth.

Kant's synthesis *a priori* is not a property of statements, or "judgments." It is an expression by which Kant foreshadowed the structuring of the phenomenon of evidence in the phenomenology of intention and fulfillment. With Husserl's contribution, the universality of the discovery of synthesis *a priori* comes into plenitude. The expression "synthesis *a priori*" does not stand for the joining of ideas, as was believed by the early interpreters of Kant; it is an expression that summarizes the fulfillment of a project of evidence-giving. What Kant discovered, and what Husserl described, is the fact that all understanding is the fulfillment of some evidence. All understanding falls under the fundamental structure of intention and fulfillment. Thanks to Kant and Husserl, we realize that all understanding is synthetic *a priori*; there is no other kind and there cannot be any other kind of understanding.

CHAPTER XV

Fundierung as a Logical Concept

Husserl's Third Logical Investigation,[1] ostensibly dealing with the phenomenology of whole and parts, is meant to introduce the notion of *Fundierung*. *Fundierung* is frequently used in phenomenological literature, although little has been written about it since Husserl introduced it. Husserl himself, while using *Fundierung* extensively, never felt the need to reopen the discussion.

Our purpose is to give a description of *Fundierung* following Husserl's original intention but using illustrative examples from the work of Wittgenstein, Ryle, and Austin, as well as Husserl.

Our method of presentation is that of phenomenological description. We do not claim to give a "pure description of the things themselves," as Husserl wanted. Phenomenological descriptions, far from being pure, are always motivated.

Accordingly, we avow from the start that we are motivated, as was Husserl, by the hope that the concept of *Fundierung* will one day enrich logic, as implication and negation have done in their time. That is, *Fundierung* is a connective which can serve as a basis for valid inferences and for the statement of necessary truths. However, *Fundierung* is not just one more trick to be added to the baggage of logic. Quite the contrary. It is likely that the adoption of the concept of *Fundierung* will alter the structure of logic more radically than Husserl might have wished.

In describing *Fundierung*, we face a difficulty similar to that of the logician who teaches the basic operations between sets, say union and intersection. A teacher has no alternative but to proceed indirectly, leading his students through a sequence of examples, hoping that the underlying concept will eventually shine through.

Our task is more difficult than the logician's. We are describing a philosophical concept which, unlike a mathematical notion, cannot be formalized in the ordinary sense; nor can we hope to introduce *Fundierung* by giving a definition. A definition, if any can be given, will be at the end rather than at the beginning.

The impossibility of formalization should not be confused with lack of rigor. Formal presentation is not the only kind of rigor. In philosophy, presentation by examples is an essential element of rigor. Examples are to philosophical discourse what logical inference is to mathematical proof. Unfortunately, the widely admired success of mathematical exposition, in which examples are informally viewed as essential but formally excluded, has worked against the acceptance of examples as part of philosophical discourse.

We begin by an example gleaned from a passage in Wittgenstein's *Philosophical Investigations*[2] which deals with reading. In this passage Wittgenstein argues that reading cannot be *reduced* to a sequence of specific acts. This conclusion is arrived at by proposing several thought experiments (what Husserl called *eidetic variations*). Each eidetic variation is followed by a counterexample. My eye movements, a detailed map of neuronal firings in the brain (obtained by the insertion of electrodes), spelling out the letters of each word one by one, understanding each word individually, and other sequences of actions are shown not to suffice to determine unequivocally that a person may be reading.

In Wittgenstein's discussion, the term "reading" is used in two senses:

(a) The *process* of reading, namely, an *event* taking place in space and time, *made up* (like a set out of elements) of "smaller" events (mental or physical) which follow each other in time.

The weight of evidence seems to support the belief that reading is a *process*. As I read, I stare at the text while moving my eyes back and forth in a characteristic manner. Neuronal changes happen in my brain as I read. These neuronal changes can be registered and plotted. It is commonly accepted that the content "on" paper is gradually being imprinted "in" my mind. The use of the prepositions "on" and "in" in this strange context is not called into question.

(b) Reading as a *function*.

I have learned the *content* of the *text* by reading the text. But logical hygiene demands that we keep the terms "text" and "content of the text" separate and equal. The text may be an object. The content of the text is not an object in any ordinary sense. Nevertheless, the content is more "important" than the text. What matters is my learning the content of the text.

My learning the content *depends* on the text and on the process of reading, but "learning the content" is not a *process* taking place at a specific where and when.

The distinctness of text and content is undeniable. It may be confirmed by eidetic variations. For example, I may learn *the same* content by reading *another* text. I may *remember* the content and forget the text.

After reading I may react to what I have read with surprise, or I may decide to make a phone call. These reactions are improperly attributed to the *process* of reading. But the content, not the text, is responsible for all my further *dealings with the world*, for determining my future course of action. True, my learning the content of the text *depends* upon the process of reading the text. Nonetheless, my further dealings with the world will be determined by the content and not by the text. It may be argued that since my further dealings with the world depend on the content, and since the content in turn "really" depends on the text, it should follow that my further dealings with the world "really" depend on the text. But the relation of "dependency," in the sense in which we are using it, is not transitive, and I am in error when I conclude that my dealings with the world depend on the text.

We come face to face with a fundamental problem: the problem of understanding what is to be meant by *dependency* of the content of what I read on the text that I read. This kind of dependency is called *Fundierung*.

Content *matters more* than text, yet content *exists less* than text. There is no way the content may be said to "exist" unless one twists the word "existence" beyond recognition. The content is not to be found anywhere. It is upsetting to accept the preceding argument, which is: what matters to us "does not exist." A great many devices have been invented to escape from facing this conclusion squarely. One such device consists of claiming that the content lies somewhere "in" the brain. Such a statement equivocates on the meaning of the preposition "in"; furthermore, it brings us face to face with an even more disquieting relationship of "dependency": the dependency of the content on the workings of the brain. Asserting that the content of what I have read "is really nothing but" neuronal firings in the brain (say) is to commit an even more grotesque error of identification than identifying the content with the text that "carries" it. Contents are not "in" the brain in any meaningful sense of the word "in."

To repeat, the conclusion is inescapable: contents do not exist anywhere, yet it is contents and not brains or texts that matter.

The next example is due to Gilbert Ryle.[3] The queen of hearts in a game of bridge is the same card (as one card in the deck of cards) as the queen of hearts in a game of poker. There is a relation of *Fundierung* between *the function* of the queen of hearts, whether in poker or in bridge, and the actual card. One cannot infer the function (the "role") of the queen of hearts in either game from plain knowledge, no matter how detailed, of the queen of hearts as a card. It is nonsensical to ask "where exactly" in the queen of hearts (as a card) can its role in a bridge game be located. Such a role is related by *Fundierung* relationships to the queen of hearts as a card as well as to brain processes, to the physics of playing cards, to the players' ambiance, and so on, *ad infinitum*.

However, the *role* of the queen of hearts in a bridge game is *nowhere*. Yet, it is the role that *matters*, not the card or the brain.

By this example we are led to another basic notion. When we focus on the content of a text, or on the role of the queen of hearts in a bridge game, we focus on a contextual *function* of the content. The *function* of the queen of hearts in a specific bridge game *matters*.

These two examples will have to suffice to lead up to the general concept. *Fundierung* is a relationship, one of whose terms is a *function*. For example, the content of the text is a *function*. This function is *related* to the text by a *Fundierung relation*. The queen of hearts as an item in a bridge game is a *function* related by a *Fundierung* relation to the queen of hearts as a card pure and simple. The other term of a *Fundierung* relation is called the *facticity*: in the first example, the text; in the second, the queen of hearts as a card in the deck.

This *relationship* between facticity and function is not reducible to any other kind of "relationship." It requires careful phenomenological description to bring out its universal occurrence. Facticity plays a "supporting role" to function. Only the function is *relevant*. The text is the *facticity* that *lets* the content *function* as relevant. We say that the content of the letter is *factically related* (or that it bears a *Fundierungsverhältnis*) to the text.

Fundierung is a *primitive relation*, one that can in no way be reduced to simpler (let alone to any "material") relations. It is the primitive logical notion that has to be admitted and understood before any experimental work on perception is undertaken. Confusing function with facticity in a *Fundierung* relation is a case of *reduction*. Reduction is the most common and devastating error of reasoning in our time. Facticity is the essential support, but it cannot upstage the function it *founds*.

Function alone is relevant. Nevertheless, function lacks autonomous standing: take away the facticity, and the function disappears with it. This tenuous umbilical cord linking relevant function to irrelevant facticity is a source of anxiety. It is hard to admit that what matters, namely functions, lacks autonomy; every effort will be made to *reduce* functions to facticities which can be observed and measured.

Psychologists and brain scientists will see to it (or so we delude ourselves) that functions are comfortingly reduced to "something concrete," something that will relieve us of the burden of admitting the lack of "existence" of "what matters."

Roles are good examples of functions. Depending on the context, I may be "playing" the roles of teacher, patient, or taxpayer. My role as a teacher is *founded* on my being a person. As functions of *Fundierung* relations, roles are structurally similar to the function of the queen of hearts in bridge.

Prices are another example of functions. The price of an item at the drugstore is given in a dollar amount. The dollar amount is the facticity of the price. However, one cannot infer the function of prices from knowledge of dollar amounts. Prices do not have any kind of existence: they are not "in" the items we buy, nor "in" our minds. Nonetheless, the "importance" of prices is beyond question. There is a *Fundierung* relation between the price of an item and the facticity of the cost expressed in dollars and cents. We cannot *reduce* prices to dollar amounts without committing a serious error of reasoning.

Tools are further striking examples of *Fundierung* relations. Pencil, paper, and ink are tools I use in writing. They are normally taken as material objects. But this is a mistake, one of many we are forced to make in our everyday dealings. Pen, paper, and ink are *functions* in *Fundierung* relations. The pen with which I write, I ordinarily take to be a material object. Strictly speaking, the pen is neither material nor object: it is a *function* that *lets* me write. I recognize this object as a pen only by virtue of my familiarity with its writing functions. The *facticities* "ink," "plastic," "small metal ball," etc., of which the pen is "made" (as we ordinarily but imprecisely say) *let* this odd-shaped object *function* as a pen. Like all facticities, they are indispensable in a pen's function; this indispensability of facticities leads to the mistaken "identification" of facticities with the function of pens. The absurdity of this reduction can be realized by eidetic variations: no amount of staring at this object as an assemblage of plastic, metal, and ink will reveal that the object we are staring at "is" a pen, unless my previous familiarity *lets* me view *the* pen *through* the facticities upon which it is *founded*.

We consider next the controversial *Fundierung* relation between viewing and seeing. This *Fundierung* relation is on a par with those of the preceding examples; however, this time we find it hard to admit that we are dealing with the same kind of relation. My viewing the pen is founded upon my seeing something; my reading the content of a printed page is founded upon my seeing it. I may see the printed page without viewing it as reading material. This may happen when the facticities of seeing intrude upon my view: I may ignore the language in which the text is written, or the printed page may be faded, etc. When reading becomes difficult or is obstructed for any reason, I stop viewing the content of the text I am reading and I view the text or the printed page instead. When I have difficulty learning the content of the text, I scan it, I try to decipher it, etc., always with a definite purpose: to *let* the facticities of the text fade so that I may *read*, that is, view the content *through* the facticities of seeing text, printed matter and other facticities.

Viewing, in manifold modes, is a *function*; seeing is the facticity that *founds* viewing. Pretending to reduce "reading" to a series of psychological or physical processes, as Wittgenstein mockingly pretends to do, is to commit the same reductionist error a child makes when he dismantles a clock to investigate the nature of time. The *Fundierung* relation separates seeing from viewing by an abyss, all the more insurmountable because it is a logical abyss. Seeing may be a process taking place in time, one that *founds* my view. But viewing has the same standing as the rules of the game of bridge, the third declension, or the cohomology of sheaves. None of these items may be said to "exist."

Lastly, let us consider the relation of dependency between the instances (the "examples") of a general concept and the concept itself. I look at the blackboard and see triangles of different shapes. One may object: I do not see triangles but imperfectly drawn shapes. Is this objection valid? We claim it is not valid. What I *view* when I stare at a drawing *of* a triangle is a triangle. The drawing *of* the triangle *founds* my view of the triangle. The misunderstanding arises out of the identification of seeing with viewing: I can only view the triangle by seeing a drawing "of it." I cannot view anything unless I see. I view the triangle through its drawings, because *all* viewing is without exception a viewing *through*. All viewing is founded upon the facticities of seeing, whether I view a content through a text, a pen through its material components, or a teacher through a person. However, simple thought experiments show that we commit the error of reduction when we "identify" viewing with seeing. Again, seeing is a physical process that can be scientifically studied from many points of view. Viewing is not a physical process. It is not even a process. It is a "happening" that does not "exist."

We are misled into reducing the function of viewing to the facticity of seeing because seeing is "real" and viewing is "imaginary." We are wedded to the prejudice that philosophy deals with "real" items, and that imaginary items are to be quietly left out. How can something imaginary matter "more" than something real?

The examples above are meant as cases *of one and the same Fundierung* relation. Our claim to the universality of *Fundierung* may not convince a skeptic any more than one can convince a skeptic that the center of a circle and a marble in a box are two instances of *the same* relation of containment between sets. There is no blinder person than one who refuses to see.

The distinction between facticity and function should be obvious from the above examples. However, the same distinction becomes difficult to admit in investigations of mental and psychological phenomena. These investigations would benefit by accepting *Fundierung* relations.

Western philosophy since the Greeks has been haunted by a *reductionist anxiety*, steadfastly refusing to draw the consequences of taking *Fundierung* seriously. The history of Western philosophy is riddled with attempts, some of them extremely clever, to reduce *Fundierung* relations to "something else" that will satisfy our cravings for certification of *existence*. We find it inadmissible that "unreal" functions should turn out to *matter*, rather than "real" objects or neurons in people's brains. Since we erroneously believe that whatever matters must *be real*, we demand that functions should be reduced to something real. Ryle's sarcasm has yet to make a dent in these demands. The firm conclusions reached by all major philosophers of this century are anathema to the fearful reductionist.

The impoverished language of set theory has armed the reductionists with one more weapon, one that is the object of Wittgenstein's jibes. As in his example, we may ask: "Is the pen a *set* (or a set of sets)? If so, *what* are the elements of each 'component' set? Are they the 'pieces' of the pen? Are plastic, ink, and metal 'elements' of the set 'pen'? Are atoms and molecules elements of some set in the chain of sets of which the pen is 'made'? Or (cribbing an example of Austin's[4]) we look at four people sitting around a table and handling small pieces of cardboard and ask, '*What* do we see? People playing bridge? People faking a bridge game? Players at a bridge championship?' "

The point is that there is no one "what" that we "see" while watching four people around a table — or while watching *anything*. All *whats* are functions in *Fundierung* relations. All *whats* "are" by the grace of some *Fundierung* relation whose *context-dependence* cannot be shoved under the rug.

The *context-dependence* of *Fundierung* is mistakenly confused by the reductionist with "arbitrariness." But this is another reductionist goof. The rules of bridge are *dependent* on the *context* of the game of bridge, but they are in no way *arbitrary*.

Wittgenstein's snapshots ridicule various tantrums that seize the reductionist who desperately avoids admitting the universality of context-dependence. Wittgenstein is aware that no amount of evidence will deter the reductionist, that he will clamor for "existence" and "reality" no matter how crushing the evidence of *Fundierung*. The thesis of phenomenology that functions neither "exist" nor "do not exist," but *are founded*, that the only kind of "existence" that makes any sense — if any — is the evanescent existence of the trump card, will only exacerbate the cravings for physical massiveness which the reductionist needs to stifle his anxiety.

After an extensive list of reductions, Wittgenstein pessimistically throws up his hands. He takes perverse pleasure in stopping short of all conclusion and goes on to the next snapshot.

CHAPTER XVI

The Primacy of Identity

To Robert Sokolowski on his sixtieth birthday

Exoteric and Esoteric Discourse

When Alfred North Whitehead wrote that all philosophy is a footnote to Greek philosophy, he did not go far enough. Our debt goes beyond philosophy. Our science, our social order, and much else we live by today can also trace their origins to the Greeks.

There is one aspect of the Greek character from which our culture has so far failed to draw full benefits: shrewdness.

The Greeks were quick to realize the dangers of allowing the public into the innermost workings of their intellectuals. With consummate shrewdness, they classified their intellectual endeavors into two kinds: the "exoteric" mysteries, open to the public, and the "esoteric" mysteries, accessible only to a few initiates.

As we range over the disciplines of our day, we note that some of them have followed the Greek example and have thereby successfully shielded the esoteric side of their subject, while opening an exoteric component to public scrutiny. Others have been reluctant to practice the Greek separation, and have suffered the consequences.

One of the most glamorous sciences is cosmology. Cosmologists are aware of the dangers of public scrutiny in areas that should remain cordoned off to technicians. The exoteric side of cosmology is a resounding success. The origin of the universe, the location of black holes in space, the existence of dark matter are live topics of conversation among cultivated people, ever eager to keep up to date with this thrilling frontier of science. Front page articles in the *New York Times* and the *Washington Post* feature the latest advances in cosmology.

Professional cosmologists are careful not to lay bare the esoteric side of their subject. The educated public might be taken aback to learn that the big bang is a bold inference drawn from a tenuous chain of tentative hypotheses, that black holes are nothing but a colorful name given to certain singularities of partial differential equations, and that dark matter is a *faute de mieux* backup.

At the other end of the scale of shrewdness we find philosophy. No discipline has contributed as much to our civilization, to finding a way out of our predicaments and to predicting our future crises. Nevertheless, the exoteric image of philosophy is one of hairsplitting argument, of gratuitous discussion of irrelevant topics. Such a dismaying dismissal stands in striking contrast with the esoteric achievements of philosophy, of which seldom have exoteric accounts been attempted.

Philosophers assume that philosophy cannot have an exoteric side, any more than it can have applications. Could it be the case that the daunting isolation of philosophy is about to come to an end? Could it be the case that the sciences of today make demands on philosophy that can no longer be ignored?

Robotics, artificial intelligence, brain science, cognitive science, in short, those sciences in which future breakthroughs are likely to be dependent upon the disentangling of foundational concepts, are becoming dependent upon philosophical argument. Whether acknowledged or not, the methods of one of the outstanding philosophical movements of this century will play a major role in the new foundational role of philosophy. I mean of course phenomenology.

The classics of phenomenology, from Husserl to Sokolowski,[1] provide a royal entry into the halls of phenomenology, and perhaps for this reason the need for presentations designed to meet the needs and objections of special audiences has not been felt. The present essay is meant as an attempt to give an exoteric presentation of one of the central ideas of phenomenology, one that recurs throughout Husserl's and Sokolowski's work.

The Phenomenology of Identity

We begin with an example.

While walking, I see a stone by the road and kick it. Sometime later, I imagine a somewhat different scene, where I walk by the same spot without kicking the stone. Years after the event, I remember the scene of my kicking the stone.

Two features of this example are relevant. In the first place, there is the *fact* that I kick, imagine, fantasize, see the *same* stone.

In the second place, when I remember the stone I kicked, my evidence for remembering *the same* stone is beyond question — in esoteric terms, it is an instance of *apodictic* evidence. Even when I walk by the same spot the next day and see the same stone in the same place, my evidence that the stone I see is the same stone I saw yesterday is unquestioned. Later findings may lead me to change my mind; yet I see the same stone. In the esoteric language of philosophers, I see the stone *as* the same.

Let us now use exoteric language. I may recall at will the event of my kicking the stone, and the stone remains the same every time I recall it. This is nothing short of miraculous. What incredible mechanism in our minds enables me to focus on the distant past with the certainty that I have recalled *the same thing*?

Before attempting an explanation of this extraordinary faculty of our minds, let us consider another example.

Every morning, I read the *New York Times*. When I pick up the rough, yellowish sheets on which the newspaper is printed, what I read is today's issue of the *New York Times*. One may argue that what I am reading is not today's *New York Times*, but a *copy* of today's *New York Times*. Digging even deeper, one may again argue that what I am reading is words on paper. Why not just black dots on a yellowish background? *What* is it then that I read when I read?

The confusion disappears when instead of using the expression "reading the *New York Times*" I use the expression "dealing with the *New York Times*." What I deal with is today's *New York Times*. There is no question about that. I confirm the fact that I am dealing with today's issue of the *New York Times*, and not with a copy of the *New York Times*, when I recognize the *same* issue of the *New York Times* on a computer screen. This recognition of identity carries the same apodictic evidence as the identity of the remembered stone.

Examples such as these can be multiplied. Anything I see, anything I think, any item I *deal with* carries the possibility of being seen, thought, or dealt with *again*. This alternation of presence and absence of the same item is not added to my act of seeing as frosting is added on the cake; quite the contrary, it is the essential component of all acts of seeing. With each and every act of seeing I *presentify* an item. In every presentification, the item maintains its identity. The stone I remember is *the same* as the stone I kicked.

Speaking exoterically, the permanence of identity through a variety of possible or actual presentifications is the constitutive property of every item.

Speaking esoterically, all physical, ideal, or psychological presentifications of any item are secondary to the one primordial phenomenon which is called *identity*. Exoterically, the world is made of objects, ideas, or whatever; esoterically, it is made of items sharing one property: their permanence throughout each presentification. Identity is the "undefined term," and the properties of identity are the axioms from which we "derive" the world.

The problem of existence is exoterically motivated by our cravings for a physical basis of the world. Esoterically, this problem is subordinate to the understanding of that fundamental happening that is the miracle of identity.

All philosophy has to pay its respects to the primacy of identity. Our exoteric slogan shall be: *"Identity precedes existence."* Esoterically, the problem of existence is a *folie*.

It is a mistake to believe that what I kick is a material stone. If the stone were a material object, then I could not remember "it," because stones do not fit in my brain. Nor does it make sense to assert that after kicking "the" stone, I retain an "image" of the stone. Such an image would have to relate to the stone I kicked, and in order to account for this "relationship" we are forced into an infinite regression. The radical realism of phenomenology consists in admitting that the stone I kick is the *very same* stone I remember. Esoterically speaking, "the stone" is neither a material object nor an imprint in my brain. "The stone" is an item that has no existence, but has an identity.

This refusal to yield to materialist or idealist reduction gives students of phenomenology the uncanny feeling that the ground they stand on is being taken away.

The Concept of Mind

The phenomenology of identity boils down to a few obvious assertions regarding "what comes first" in the composition of the world and in our perception of it. Like all philosophical doctrines, the tenets of phenomenology are as trivial as the theorems of mathematics. Nevertheless, philosophers and scientists steeped in the empiricist tradition find it difficult to face the consequences that accrue to admitting the primacy of identity over existence.

The restructuring of the concept of mind will be drastic. Those extraordinary faculties that make it possible for the mind to perceive identity are extraordinary only to those who are wedded to mechanistic preconceptions.

Phenomenology proposes an inversion of priorities in the concept of mind. In the empirical tradition the mechanistic functions of the mind are assumed to be fundamental while the perception of identity is taken as a secondary function. Phenomenology asserts instead that the converse is true.

There is no way to "reduce" identity to any mental process because it is absurd to reduce identity to a "process." No amount of scientific research will succeed in "explaining" the mechanism of identity because identity is not a mechanism. What has been called "the presupposition error" consists of ignoring the fact that scientific investigations of the brain presuppose the phenomenon of identity. This presupposition is seldom acknowledged.

A great many of the scientific "models" of the brain that have been proposed fall into the presupposition error.

We find a typical instance of the presupposition error in a recent article by the neurologist Steven Rose. He writes:

> The ultimate problem — which has been called the binding problem — is *how* the modules [of the mind] are integrated so each of us gets an apparently seamless conscious experience. The problem is analogous to the way in a movie theater we see a series of rapidly projected discrete pictures as a smoothly moving image. To this question, no [one] can yet offer an answer — and this is not surprising, because it is still *the* greatest question for everyone working at the interface of mind and brain.[2]

What the author calls "the binding problem" is the presupposition of identity. The author assumes that the phenomenon described as "a series of rapidly projected discrete pictures [seen] as a smoothly moving image" is obvious, and thereby misses the philosophical problem of the conditions of possibility of *identical* images. He might be taken aback to learn that without an account of such conditions of possibility, no progress will be made in explaining how discrete pictures can be seen as a smoothly moving image. In the unphilosophical age from which we are emerging, such an assertion is likely to be either missed or dismissed.

CHAPTER XVII

Three Senses of "A is B" in Heidegger

It may be helpful to subdivide Heidegger's view of "A is B" into three stages. However, such a subdivision is no more than a rhetorical device. The development of a person's thought refuses linearization.

The tradition of philosophy constrains us to use words like "problems," "solutions," "arguments," and "relationship." There is at present no alternative to this language. Heidegger attempted to develop a language which he considered more appropriate, and time will tell whether his lead can be followed.

The first stage of Heidegger's reflection on language[1] is contemporary with the forward leap in logic that began with the work of Frege and the symbolic notation invented by Peano at the turn of the century. Young Heidegger's book reviews shared the enthusiasm for the new logic. At last logic could claim to have risen again to the level of Duns Scotus.

By accident, Heidegger received his training in scholastic logic. Even if Heidegger had been a student of Frege, his critique of logic would not be different. Scholastic logic and mathematical logic have similar objectives. They are logics of objectivity, founded on those assumptions that the later Heidegger summarized in the word *Gestell.*

Together with the success of the new logic came its end. As the program of the predicate calculus was completed, the hidden assumptions lying beneath formal logic became apparent. The program of formalization that began with Aristotle and that concluded with Frege ignores the ur-problem of the *copula*, namely, the problem of the meaning of the word *is* in the statement "*A* is *B*." Short of assuming the identity of *A* and *B*, an assumption that is belied by our use of two distinct letters, "*A*" and "*B*," it is inconceivable how "A" can *be* something other than itself.

The last philosopher to give serious thought to this problem prior to Heidegger was Hegel. Heidegger, however, chose not to deal with the heritage of Hegel, and followed instead the method of phenomenology inaugurated by Husserl. Thus began the second stage of Heidegger's reflection on "A is B."

Husserl's philosophy may be labelled "*salus ex descriptione.*" Husserl was aware of the problem of the copula, but with characteristic reluctance he does not stage a frontal assault. Instead, he uses the problem as a motivation for a milder critique of logic.

After its symbolic rewriting in the language of set theory, logic reduced the interpretations of *is* in the sentence "*A* is *B*" to one of only two senses: either as "*A* is contained in *B*" or as "*A* is a member of *B*," where *A* and *B* are sets. Husserl uncovered other relationships between *A* and *B* which he proved not to be reducible to containment or membership of sets. He laid bare the underlying assumptions of formal logic by showing that logic gives favored treatment to two relations while ignoring all other relations, some of which deserve to be made rigorous. It was Husserl's intention to enlarge logic by *formalizing* new relationships, though it is not clear what Husserl intended by "formalization."

Heidegger took advantage of Husserl's spadework[2] to develop a new language designed for an attack on the foundational problems of logic. He made use of Husserl's discovery of new relations in a more radical way than his teacher might have wished. One such relation is singled out as fundamental: the relation of *Fundierung*. In his "Third Logical Investigation," Husserl had foreshadowed the central role of *Fundierung*, but he did not go far enough in implementing his discovery.

Heidegger's argument leads to the dissolution of the problem of the copula as well as to the discovery of the conditions of possibility of *Fundierung*. A stereotyped summary might go as follows.

As I speak to you, you listen to me through sounds which somehow make sense. The sense of my words is not the same as the sound. Sense and sound are related by a *Fundierungszusammenhang*. Husserl gave an incisive description of *Fundierungszusammenhang*, the upshot of which is to prove that the relations of membership and containment are subordinate to the primitive relation of *Fundierung*.

Heidegger shows that the the conditions of possibility of *Fundierung* can be found only if *Fundierung* is viewed as a special (ontic) instance of some universal (ontological) problem. The *Zusammenhang* between *A* and *B* is a problem only when the phenomena of *Fundierung* and "A is B" are taken in isolation. The problem is made to disappear by uncovering a universal notion of which both *Fundierung* as well as the "is" in "A is B" will be special instances. Where shall we find the betokening of such universality? We will find it when we place *Fundierung* and the "is" in parallel with other phenomena which will all be seen as instances of the one and the same universal concept.

Descartes had seen the problem of the *is* in "*A* is *B*" as an impenetrable mystery. Heidegger does not attempt to dispel the mystery. Instead, he shows that Descartes' mystery is "the same" mystery as several others, for example, the mystery of *Fundierung*. He shows that "the same" mystery is found in all speech whatsoever. You see a mystery in the *is* of "*A* is *B*," you see another mystery in *Fundierung*, because of the prejudice that the two mysteries of the *is* and of the "*Fundierungszusammenhang*" are of a different kind. But what if we realized that the same mystery is the condition of possibility of *all* "relationships?" Then the mysteries would reduce to one single mystery, namely, the condition of possibility of "relationships." But a universal mystery is no different from a universal law. Heidegger concludes with the discovery of the universal law of the *as*.

The universal *as* is given various names in Heidegger's writings: it will be the primordial *Nicht* between beings and Being, the ontological difference, the Beyond, the primordial *es gibt*, the *Ereignis*.

The discovery of the universal *as* is Heidegger's contribution to philosophy.[3]

Heidegger's later thought in no way alters this discovery. He came to believe that the language of phenomenology, in which his middle writings are couched, was inadequate to his discovery. The later Heidegger wanted to recast his discovery in a non-objectivistic language, since the universal *as* lies beyond objectivity. The universal "as" is the surgence of sense in Man, the shepherd of Being.

The disclosure of the primordial *as* is the end of a search that began with Plato, followed a long route through Descartes, Leibniz, Kant, Vico, Hegel, Dilthey and Husserl. This search comes to its conclusion with Heidegger.

PART III

Indiscrete Thoughts

CHAPTER XVIII

Ten Lessons I wish I Had Been Taught

MIT, April 20 , 1996 on the occasion of the Rotafest

ALLOW ME TO BEGIN by allaying one of your worries. I will not spend the next half hour thanking you for participating in this conference, or for your taking time away from work to travel to Cambridge.

And to allay another of your probable worries, let me add that you are not about to be subjected to a recollection of past events similar to the ones I've been publishing for some years, with a straight face and an occasional embellishment of reality.

Having discarded these two choices for this talk, I was left without a title. Luckily I remembered an MIT colloquium that took place in the late fifties; it was one of the first I attended at MIT. The speaker was Eugenio Calabi. Sitting in the front row of the audience were Norbert Wiener, asleep as usual until the time came to applaud, and Dirk Struik who had been one of Calabi's teachers when Calabi was an undergraduate at MIT in the forties. The subject of the lecture was beyond my competence. After the first five minutes I was completely lost. At the end of the lecture, an arcane dialogue took place between the speaker and some members of the audience, Ambrose and Singer if I remember correctly. There followed a period of tense silence. Professor Struik broke the ice. He raised his hand and said: "Give us something to take home!" Calabi obliged, and in the next five minutes he explained in beautiful simple terms the gist of his lecture. Everybody filed out with a feeling of satisfaction.

Dirk Struik was right: a speaker should try to give his audience something they can take home. But what? I have been collecting some random bits of advice that I keep repeating to myself, do's and don'ts of which I have been and will always be guilty. Some of you have been exposed to one or more of these tidbits. Collecting these items and presenting them in one speech may be one of the less obnoxious among options of equal presumptuousness. The advice we give others is the advice that we ourselves need. Since it is too late for me to learn these lessons, I will discharge my unfulfilled duty by dishing them out to you. They will be stated in order of increasing controversiality.

1. Lecturing

The following four requirements of a good lecture do not seem to be altogether obvious, judging from the mathematics lectures I have been listening to for the past forty-six years.

a. Every lecture should make only one main point

The German philosopher G. W. F. Hegel wrote that any philosopher who uses the word "and" too often cannot be a good philosopher. I think he was right, at least insofar as lecturing goes. Every lecture should state one main point and repeat it over and over, like a theme with variations. An audience is like a herd of cows, moving slowly in the direction they are being driven towards. If we make one point, we have a good chance that the audience will take the right direction; if we make several points, then the cows will scatter all over the field. The audience will lose interest and everyone will go back to the thoughts they interrupted in order to come to our lecture.

b. Never run overtime

Running overtime is the one unforgivable error a lecturer can make. After fifty minutes (one microcentury as von Neumann used to say) everybody's attention will turn elsewhere even if we are trying to prove the Riemann hypothesis. One minute overtime can destroy the best of lectures.

c. Relate to your audience

As you enter the lecture hall, try to spot someone in the audience with whose work you have some familiarity. Quickly rearrange your presentation so as to manage to mention some of that person's work. In this way, you will guarantee that at least one person will follow with rapt attention, and you will make a friend to boot.

Everyone in the audience has come to listen to your lecture with the secret hope of hearing their work mentioned.

d. Give them something to take home

It is not easy to follow Professor Struik's advice. It is easier to state what features of a lecture the audience will always remember, and the answer is not pretty.

I often meet, in airports, in the street and occasionally in embarrassing situations, MIT alumni who have taken one or more courses from me. Most of the time they admit that they have forgotten the subject of the course, and all the mathematics I thought I had taught them. However, they will gladly recall some joke, some anecdote, some quirk, some side remark, or some mistake I made.

2. Blackboard Technique

Two points.

a. Make sure the blackboard is spotless

It is particularly important to erase those distracting whirls that are left when we run the eraser over the blackboard in a non uniform fashion.

By starting with a spotless blackboard, you will subtly convey the impression that the lecture they are about to hear is equally spotless.

b. Start writing on the top left hand corner

What we write on the blackboard should correspond to what we want an attentive listener to take down in his notebook. It is preferable to write slowly and in a large handwriting, with no abbreviations. Those members of the audience who are taking notes are doing us a favor, and it is up to us to help them with their copying. When slides are used instead of the blackboard, the speaker should spend some time explaining each slide, preferably by adding sentences that are inessential, repetitive or superfluous, so as to allow any member of the audience time to copy our slide. We all fall prey to the illusion that a listener will find the time to read the copy of the slides we hand them after the lecture. This is wishful thinking.

3. Publish the same result several times

After getting my degree, I worked for a few years in functional analysis. I bought a copy of Frederick Riesz' *Collected Papers* as soon as the big thick heavy oversize volume was published. However, as I began to leaf through, I could not help but notice that the pages were extra thick, almost like cardboard. Strangely, each of Riesz' publications had been reset in exceptionally large type. I was fond of Riesz' papers, which were invariably beautifully written and gave the reader a feeling of definitiveness.

As I looked through his *Collected Papers* however, another picture emerged. The editors had gone out of their way to publish every little scrap Riesz had ever published. It was clear that Riesz' publications were few. What is more surprising is that the papers had been published several times. Riesz would publish the first rough version of an idea in some obscure Hungarian journal. A few years later, he would send a series of notes to the French Academy's *Comptes Rendus* in which the same material was further elaborated. A few more years would pass, and he would publish the definitive paper, either in French or in English.

Adam Koranyi, who took courses with Frederick Riesz, told me that Riesz would lecture on the same subject year after year, while meditating on the definitive version to be written. No wonder the final version was perfect.

Riesz' example is worth following. The mathematical community is split into small groups, each one with its own customs, notation and terminology. It may soon be indispensable to present the same result in several versions, each one accessible to a specific group; the price one might have to pay otherwise is to have our work rediscovered by someone who uses a different language and notation, and who will rightly claim it as his own.

4. You are more likely to be remembered by your expository work

Let us look at two examples, beginning with Hilbert. When we think of Hilbert, we think of a few of his great theorems, like his basis theorem. But Hilbert's name is more often remembered for his work in number theory, his *Zahlbericht*, his book *Foundations of Geometry* and for his text on integral equations. The term "Hilbert space" was introduced by Stone and von Neumann in recognition of Hilbert's textbook on integral equations, in which the word "spectrum" was first defined at least twenty years before the discovery of quantum mechanics. Hilbert's textbook on integral equations is in large part expository, leaning on the work of Hellinger and several other mathematicians whose names are now forgotten.

Similarly, Hilbert's *Foundations of Geometry*, the book that made Hilbert's name a household word among mathematicians, contains little original work, and reaps the harvest of the work of several geometers, such as Kohn, Schur (not the Schur you have heard of), Wiener (another Wiener), Pasch, Pieri and several other Italians.

Again, Hilbert's *Zahlbericht*, a fundamental contribution that revolutionized the field of number theory, was originally a survey that Hilbert was commissioned to write for publication in the Bulletin of the German Mathematical Society.

William Feller is another example. Feller is remembered as the author of the most successful treatise on probability ever written. Few probabilists of our day are able to cite more than a couple of Feller's research papers; most mathematicians are not even aware that Feller had a previous life in convex geometry.

Allow me to digress with a personal reminiscence. I sometimes publish in a branch of philosophy called phenomenology. After publishing my first paper in this subject, I felt deeply hurt when, at a meeting of the Society for Phenomenology and Existential Philosophy, I was rudely told in no uncertain terms that everything I wrote in my paper was well known. This scenario occurred more than once, and I was eventually forced to reconsider my publishing standards in phenomenology.

It so happens that the fundamental treatises of phenomenology are written in thick, heavy philosophical German. Tradition demands that no examples ever be given of what one is talking about. One day I decided, not without serious misgivings, to publish a paper that was essentially an updating of some paragraphs from a book by Edmund Husserl, with a few examples added. While I was waiting for the worst at the next meeting of the Society for Phenomenology and Existential Philosophy, a prominent phenomenologist rushed towards me with a smile on his face. He was full of praise for my paper, and he strongly encouraged me to further develop the novel and original ideas presented in it.

5. Every mathematician has only a few tricks

A long time ago an older and well known number theorist made some disparaging remarks about Paul Erdös' work. You admire Erdös' contributions to mathematics as much as I do, and I felt annoyed when the older mathematician flatly and definitively stated that all of Erdös' work could be "reduced" to a few tricks which Erdös repeatedly relied on in his proofs. What the number theorist did not realize is that other mathematicians, even the very best, also rely on a few tricks which they use over and over. Take Hilbert. The second volume of Hilbert's collected papers contains Hilbert's papers in invariant theory. I have made a point of reading some of these papers with care. It is sad to note that some of Hilbert's beautiful results have been completely forgotten. But on reading the proofs of Hilbert's striking and deep theorems in invariant theory, it was surprising to verify that Hilbert's proofs relied on the same few tricks. Even Hilbert had only a few tricks!

6. Do not worry about your mistakes

Once more let me begin with Hilbert. When the Germans were planning to publish Hilbert's collected papers and to present him with a set on the occasion of one of his later birthdays, they realized that they could not publish the papers in their original versions because they were full of errors, some of them quite serious. Thereupon they hired a young unemployed mathematician, Olga Taussky-Todd, to go over Hilbert's papers and correct all mistakes. Olga labored for three years; it turned out that all mistakes could be corrected without any major changes in the statement of the theorems. There was one exception, a paper Hilbert wrote in his old age, which could not be fixed; it was a purported proof of the continuum hypothesis, you will find it in a volume of the *Mathematische Annalen* of the early thirties. At last, on Hilbert's birthday, a freshly printed set of Hilbert's collected papers was presented to the *Geheimrat*. Hilbert leafed through them carefully and did not notice anything.

Now let us shift to the other end of the spectrum, and allow me to relate another personal anecdote. In the summer of 1979, while attending a philosophy meeting in Pittsburgh, I was struck with a case of detached retinas. Thanks to Joni's prompt intervention, I managed to be operated on in the nick of time and my eyesight was saved.

On the morning after the operation, while I was lying on a hospital bed with my eyes bandaged, Joni dropped in to visit. Since I was to remain in that Pittsburgh hospital for at least a week, we decided to write a paper. Joni fished a manuscript out of my suitcase, and I mentioned to her that the text had a few mistakes which she could help me fix.

There followed twenty minutes of silence while she went through the draft. "Why, it is all wrong!" she finally remarked in her youthful voice. She was right. Every statement in the manuscript had something wrong. Nevertheless, after laboring for a while, she managed to correct every mistake, and the paper was eventually published.

There are two kinds of mistakes. There are fatal mistakes that destroy a theory; but there are also contingent ones, which are useful in testing the stability of a theory.

7. Use the Feynman method

Richard Feynman was fond of giving the following advice on how to be a genius. You have to keep a dozen of your favorite problems constantly present in your mind, although by and large they will lay in a dormant state. Every time you hear or read a new trick or a new result, test it against each of your twelve problems to see whether it helps. Every once in a while there will be a hit, and people will say: "How did he do it? He must be a genius!"

8. Give lavish acknowledgments

I have always felt miffed after reading a paper in which I felt I was not being given proper credit, and it is safe to conjecture that the same happens to everyone else. One day, I tried an experiment. After writing a rather long paper, I began to draft a thorough bibliography. On the spur of the moment, I decided to cite a few papers which had nothing whatsoever to do with the content of my paper, to see what might happen.

Somewhat to my surprise, I received letters from two of the authors whose papers I believed were irrelevant to my article. Both letters were written in an emotionally charged tone. Each of the authors warmly congratulated me for being the first to acknowledge their contribution to the field.

9. Write informative introductions

Nowadays, reading a mathematics paper from top to bottom is a rare event. If we wish our paper to be read, we had better provide our prospective readers with strong motivation to do so. A lengthy introduction, summarizing the history of the subject, giving everybody his due, and perhaps enticingly outlining the content of the paper in a discursive manner, will go some of the way towards getting us a couple of readers.

As the editor of the journal *Advances in Mathematics*, I have often sent submitted papers back to the authors with the recommendation that they lengthen their introduction. On occasion I received by return mail a message from the author, stating that the same paper had been previously rejected by *Annals of Mathematics* because the introduction was already too long.

10. Be prepared for old age

My late friend Stan Ulam used to remark that his life was sharply divided into two halves. In the first half, he was always the youngest person in the group; in the second half, he was always the oldest. There was no transitional period.

I now realize how right he was. The etiquette of old age does not seem to have been written up, and we have to learn it the hard way. It depends on a basic realization, which takes time to adjust to. You must realize that, after reaching a certain age, you are no longer viewed as a person. You become an institution, and you are treated the way institutions are treated. You are expected to behave like a piece of period furniture, an architectural landmark, or an incunabulum.

It matters little whether you keep publishing or not. If your papers are no good, they will say, "What did you expect? He is a fixture!" and if an occasional paper of yours is found to be interesting, they will say, "What did you expect? He has been working at this all his life!" The only sensible response is to enjoy playing your newly-found role as an institution.

CHAPTER XIX

Ten Lessons for the Survival of a Mathematics Department

Times are changing: mathematics, once the queen of the sciences and the undisputed recipient of research funds, is now being shoved aside in favor of fields which are (wrongly) presumed to have applications, either because they endow themselves with a catchy terminology, or because they know (better than mathematicians ever did) how to make use of the latest techniques in P.R. The following decalogue was written as a message of warning to a colleague who insisted that all is well and that nothing can happen to us mathematicians as long as we keep proving deep theorems.

1. Never wash your dirty linen in public

I know that you frequently (and loudly if I may add) disagree with your colleagues about the relative value of fields of mathematics and about the talents of practicing mathematicians. All of us hold some of our colleagues in low esteem, and sometimes we cannot keep ourselves from sharing these opinions with our fellow mathematicians.

When talking to your colleagues in *other* departments, however, these opinions should never be brought up. It is a mistake for you to think that you might thereby gather support against mathematicians you do not like. What your colleagues in other departments will do instead, after listening to you, is use your statements as proof of the weakness of the whole mathematics department, to increase their own department's standing at the expense of mathematics.

Departments of a university are like sovereign states: there is no such thing as charity towards one another.

2. Never go above the head of your department

When a dean or a provost receives a letter from a distinguished faculty member like you which ignores your chairman's opinion, his or her reaction is likely to be one of irritation. It matters little what the content of the letter might be. You see, the letter you have sent forces him or her to think about matters that he or she thought should be dealt with by the chairman of the department. Your letter will be viewed as evidence of disunity in the rank and file of mathematicians.

Human nature being what it is, such a dean or provost is likely to remember an unsolicited letter at budget time, and not very kindly at that.

3. Never compare fields

You are not alone in believing that your own field is better and more promising than those of your colleagues. We all believe the same about our own fields. But our beliefs cancel each other out. Better keep your mouth shut rather than make yourself obnoxious. And remember, when talking to outsiders, have nothing but praise for your colleagues in all fields, even for those in combinatorics. All public shows of disunity are ultimately harmful to the well-being of mathematics.

4. Remember that the grocery bill is a piece of mathematics too

Once, during a year at a liberal arts college, I was assigned to teach a course on "Mickey Mouse math." I was stung by a colleague's remark that the course "did not deal with real mathematics." It certainly wasn't a course in physics or chemistry, was it?

We tend to use the word "mathematics" in a valuative sense, to denote the kind of mathematics we and our friends do. This is a mistake. The word "mathematics" is correctly used in a strictly objective sense. The grocery bill, a computer program, and class field theory are three instances of mathematics. Your opinion that some instances may be better than others is most effectively verbalized when you are asked to vote on a tenure decision. At other times, a careless statement of relative values is more likely to turn potential friends of mathematics into enemies of our field. Believe me, we are going to need all the friends we can get.

5. Do not look down on good teachers

Mathematics is the greatest undertaking of mankind. All mathematicians know this. Yet many people do not share this view. Consequently, mathematics is not as self-supporting a profession in our society as the exercise of poetry was in medieval Ireland. Most of our income will have to come from teaching, and the more students we teach, the more of our friends we can appoint to our department. Those few colleagues who are successful at teaching undergraduate courses should earn our thanks as well as our respect. It is counterproductive to turn up our noses at those who bring home the dough.

When Mr. Smith dies and decides to leave his fortune to our mathematics department, it will be because he remembers his good teacher Dr. Jones who never made it beyond associate professor, not because of the wonderful research papers you have written.

6. Write expository papers

When I was in graduate school, one of my teachers told me, "When you write a research paper, you are afraid that your result might already be known; but when you write an expository paper, you discover that nothing is known."

Not only is it good for you to write an expository paper once in a while, but such writing is essential for the survival of mathematics. Look at the most influential writings in mathematics of the last hundred years. At least half of them would have to be classified as expository.

Let me put it to you in the P.R. language that you detest. It is not enough for you (or anyone) to have a good product to sell; you must package it right and advertise it properly. Otherwise you will go out of business.

Now don't tell me that you are a pure mathematician and therefore that you are above and beyond such lowly details. It is the results of pure mathematics and not of applied mathematics that are most sought-after by physicists and engineers (and soon, we hope, by biologists as well). Let us do our best to make our results available to them in a language they can understand. If we don't, they will some day no longer believe we have any new results, and they will cut off our research funds. Remember, they are the ones who control the purse strings since we mathematicians have always proven ourselves inept in all political and financial matters.

7. Do not show your questioners to the door

When an engineer knocks at your door with a mathematical question, you should not try to get rid of him or her as quickly as possible. You are likely to make a mistake I myself made for many years: to believe that the engineer wants you to solve his or her problem. This is the kind of oversimplification for which we mathematicians are notorious. Believe me, the engineer does not want you to solve his or her problem. Once, I did so by mistake (actually, I had read the solution in the library two hours previously, quite by accident) and he got quite furious, as if I were taking away his livelihood. What the engineer wants is to be treated with respect and consideration, like the human being he is, and most of all to be listened to with rapt attention. If you do this, he will be likely to hit upon a clever new idea as he explains the problem to you, and you will get some of the credit.

Listening to engineers and other scientists is our duty. You may even learn some interesting new mathematics while doing so.

8. View the mathematical community as a United Front

Grade school teachers, high school teachers, administrators and lobbyists are as much mathematicians as you or Hilbert. It is not up to us to make invidious distinctions. They contribute to the well-being of mathematics as much as or more than you or other mathematicians. They are right in feeling left out by snobbish research mathematicians who do not know on which side their bread is buttered. It is our best interest, as well as the interest of justice, to treat all who deal with mathematics in whatever way as equals. By being united we will increase the probability of our survival.

9. Attack flakiness

Now that Communism is a dead duck, we need a new Threat. Remember, Congress only reacts to potential or actual threats (through no fault of their own, it is the way the system works). Flakiness is nowadays creeping into the sciences like a virus through a computer, and it may be the present threat to our civilization. Mathematics can save the world from the invasion of the flakes by unmasking them and by contributing some hard thinking. You and I know that mathematics is not and will never be flaky, by definition.

This is the biggest chance we have had in a long while to make a lasting contribution to the well-being of Science. Let us not botch it as we did with the few other chances we have had in the past.

10. Learn when to withdraw

Let me confess to you something I have told very few others (after all, this message will not get around much): I have written some of the papers I like the most while hiding in a closet. When the going gets rough, we have recourse to a way of salvation that is not available to ordinary mortals: we have that Mighty Fortress that is our Mathematics. This is what makes us mathematicians into very special people. The danger is envy from the rest of the world.

When you meet someone who does not know how to differentiate and integrate, be kind, gentle, understanding. Remember, there are lots of people like that out there, and if we are not careful, they will do away with us, as has happened many times before in history to other Very Special People.

CHAPTER XX

A Mathematician's Gossip

A good teacher does not teach facts, he or she teaches enthusiasm, open-mindedness and values. Young people need encouragement. Left to themselves, they may not know how to decide what is worthwhile. They may drop an original idea because they think someone else must have thought of it already. Students need to be taught to believe in themselves and not to give up.

Von Neumann used to say that a mathematician is finished by the age of thirty. As he got older, he increased the age to thirty-five, then to forty, to forty-five, and soon to fifty. We have inherited from the 19th century the prejudice that mathematicians have to do their work early before they are finished. This is not true. The kind of work a mathematician does as he grows older changes. An older mathematician will work on questions of wider scope, whereas a younger one may choose to work on a single difficult problem.

There is a ratio by which you can measure how good a mathematician is, and that is how many crackpot ideas he must have in order to have one good one. If this ratio equals ten to one then he is a genius. For the average mathematician, it may be one hundred to one. You must have the courage to discard attractive ideas. This is a feature of creative thinking that the man in the street fails to appreciate.

What makes a mathematician creative? Rule one: don't ask him to be creative. There is nothing deadlier for a mathematician than to be placed in a beautiful office and instructed to lay golden eggs. Creativity is never directly sought after. It comes indirectly. It comes while you are complaining about too much routine work, and you decide to spend half an hour on your project. Or while getting ready to lecture, you realize that the textbook is lousy, and that the subject has never been properly explained. While you work at explaining some old material, lo and behold, you get a great new idea. Creativity is a bad word. Unfortunately, we must leave it in the books because people in power believe in it. It is a dangerous word.

A friend of mine, a well known painter, was looking at a painting by Velázquez. I watched her reactions. She started by saying, "How funny, this stroke is going down! Normally, we brush this way, but he brushed that way." Then, "This is a combination of colors I have never seen." She said nothing about the artist's creativity. It is demoralizing to hold up Einstein or Beethoven as examples of creativity to be imitated by children. The idea of genius, elaborated by German romantics, is destructive; it is a flight into fantasy, like Nazism. There is reason to believe we have killed classical music because of our prejudices on genius and creativity. People are browbeaten into believing that unless they can be geniuses like Beethoven they might as well quit. But look at the Baroque Age — there were hundreds of little Italians who wrote good music and did not give a hoot about being creative geniuses.

The Talmud and the Tao are trainers of scientific minds.

A variety of talents is required if the scientific community is to thrive. The farthest reaching contribution of psychology to human welfare is the realization that intelligence is not a monolithic faculty that can be measured on a linear scale. You may be smart at math problems, but stupid at everything else. The IQ test was effective at testing one kind of intelligence, what we might call quiz-kid smarts, but it did not test any of the other kinds.

The eye-catching, entertainment-minded, ornamental use of mathematical jargon in subjects which are still ages away from mathematization is an embarrassing legacy of the irresponsible sixties. It has done irreparable harm to mathematics and has caused reactions which may go too far in their condemnation.

The history of mathematics is replete with injustice. There is a tendency to exhibit toward the past a forgetful, oversimplifying, hero-worshiping attitude that we have come to identify with mass behavior. Great advances in science are pinned on a few extraordinary white-maned individuals. By the magic powers of genius denied to ordinary mortals (thus safely getting us off the hook), they alone are made responsible for Progress.

The public abhors detail. Revealing that behind every great man one can find a beehive of lesser-known individuals who paved his way and obtained most of the results for which he is known is a crime of *lèse majesté*. Whoever dares associate Apollonius with Euclid, Cavalieri with Leibniz, Saccheri with Lobachevski, Kohn with Hilbert, MacMahon with Ramanujan should stand ready for the scornful reaction of the disappointed majority.

One consequence of this sociological law is that whenever a forgotten branch of mathematics comes back into fashion after a period of neglect only the main outlines of the theory are remembered, those you would find in the works of the Great Men. The bulk of the theory is likely to be rediscovered from scratch by smart young mathematicians who have realized that their future careers depend on publishing research papers rather than on rummaging through dusty old journals.

In all mathematics, it would be hard to find a more blatant instance of this regrettable state of affairs than the theory of symmetric functions. Each generation rediscovers them and presents them in the latest jargon. Today it is K-theory, yesterday it was categories and functors, and the day before, group representations. Behind these and several other attractive theories stands one immutable source: the ordinary, crude definition of the symmetric functions and the identities they satisfy.

"*Historia magistra vitae, lux veritatis...*," we repeat to ourselves without conviction, and would like to go on believing. But scientific and technological history seem to belie this saying. The early version of an as yet ill-understood algorithm, the clumsy parts of an early engine, the pristine computer with its huge, superfluous circuits give us little inspiration to face the problems of our day. Technological advances appear as sudden, discontinuous leaps that cover all previous work with an impenetrable cobweb of obsolescence. It is left to the archeologist, not the historian to make way across the maze of oblivion and to retrieve an appearance of the lost artifact, an obtrusive contraption whose plans and photographs will fill the glossy pages of coffee table books.

Or so we are tempted to think when we look with secret boredom at the Carrollesque creations of Babbage, at the megalomaniac plans of Geheimrat Leibniz, at the unconvincing fantasies of Leonardo da Vinci, or at the preposterous wheels of Raimond Lull. There is a point at which the study of the technological past turns into paleontology, and in the history of computation that point is uncomfortably close and moving closer.

It is hard to tell the crackpots from the geniuses. The same person may be both, as in the beginnings of science when Kepler identified the distances of the planets from the sun with the lengths of the sides of the five regular solids inscribed in a sphere. Newton believed in magic and nevertheless discovered the laws of mechanics. The new sciences are a melting pot of good guys and bad guys, of con men and serious people. It is hard to sort them out.

Most people, even some scientists, think that mathematics applies because you learn Theorem Three and Theorem Three somehow explains the laws of nature. This does not happen even in science fiction novels; it is pure fantasy. The results of mathematics are seldom directly applied; it is the definitions that are really useful. Once you learn the concept of a differential equation, you see differential equations all over, no matter what you do. This you cannot see unless you take a course in *abstract* differential equations. What applies is the cultural background you get from a course in differential equation, not the specific theorems. If you want to learn French, you have to live the life of France, not just memorize thousands of words. If you want to apply mathematics, you have to live the life of differential equations. When you live this life, you can then go back to molecular biology with a new set of eyes that will see things you could not otherwise see.

I once entertained the thought that biologists could tell me what their mathematical problems were, so that I could solve them. This was ridiculous. Biologists seldom have the mathematical view that is required to spot problems in the mathematics of biology that are staring at them. A biologist will never see anything deeper than binomial coefficients. It is not that the problems aren't there; rather, biologists don't have the view that comes with a solid education in mathematics.

One of the rarest mathematical talents is the talent for applied mathematics, for picking out of a maze of experimental data the two or three parameters that are relevant, and to discard all other data. This talent is rare. It is taught only at the shop level.

Nature imitates mathematics.

Very little mathematics has direct applications — though fortunately most of it has plenty of indirect ones.

There are too many revolutions going on, perhaps because none of them is real enough to be called a revolution. The computer revolution is fizzling out faster than could have been predicted. Will the information revolution be next?

Mathematics is the study of analogies between analogies. All science is. Scientists want to show that things that don't look alike are really the same. That is one of their innermost Freudian motivations. In fact, that is what we mean by understanding.

The apex of mathematical achievement occurs when two or more fields which were thought to be entirely unrelated turn out to be closely intertwined. Mathematicians have never decided whether they should feel excited or upset by such events.

The progress of mathematics may be viewed as a movement from the infinite to the finite. At the start, the possibilities of a theory, for example the theory of enumeration, appear boundless. Rules for the enumeration of sets subject to various conditions appear to obey an indefinite variety of recursions and seem to lead to a bounty of generating functions. We are naively led to conjecture that the class of enumerable objects is infinite and unclassifiable.

As cases pile upon cases, however, patterns begin to emerge. Freakish instances are quietly disregarded; impossible problems are recognized as such, and what is left gets organized along a few general criteria. We will do all we can to boil these criteria down to one, but we will probably have to be satisfied with a small finite number.

"Mainstream mathematics" is a name given to the mathematics that more fittingly belongs on Sunset Boulevard.

Milan has succeeded in overtaking Paris in women's fashions, but Paris retains the title for intellectual elegance. The finest presentations of today's mathematics are heard in the halls of the Institut Henri Poincaré, where polite insults are elegantly exchanged with jewels of mathematical definitiveness.

The French are remarkable for the stability of the educational institutions they have created in which the life of intellect can thrive: the *Collège de France*, the *Grandes Écoles*, the *Institut*, and. . . the *Séminaire Bourbaki*! It is, among all organizations meant to keep mathematics together, the most successful, the most admired and the most envied. *Vive la France!*

Mathematicians have to attend (secretly) physics meetings in order to find out what is going on in their fields. Physicists have the P.R., the savoir-faire, and the chutzpah to write readable, or at least legible accounts of subjects that are not yet obsolete, something few mathematicians would dare to do, fearing expulsion from the A.M.S.

A common error of judgment among mathematicians is the confusion between telling the truth and giving a logically correct presentation. The two objectives are antithetical and hard to reconcile. Most presentations obeying the current *Diktats* of linear rigor are a long way from telling the truth; any reader of such a presentation is forced to start writing on the margin, or deciphering on a separate sheet of paper.

The truth of any piece of mathematical writing consists of realizing what the author is "up to"; it is the tradition of mathematics to do whatever it takes to avoid giving away this secret. When an author lets the truth slip out, the accusation of being "sloppy," "philosophical," "digressing," or worse, is instantly made.

Some theorems are hygienic prescriptions meant to guard us against potentially unpleasant complications. Authors of mathematics books frequently forget to give any hint as to what these complications will be. This omission makes their exposition incomprehensible. This equivocation is common in commutative algebra, where theorems sounding like "All extremely regular rings are fully normal" are proved with an abundance of mysterious detail and followed by pointless examples of extremely regular and fully normal rings. Authors of algebra textbooks fail to realize that hygienic theorems can only be understood by giving examples of rings that are *not* extremely regular and *not* fully normal.

Some subjects can be roughly associated with geographic locations: graph theory is a Canadian subject, singular integrals is an Argentine subject, class field theory an Austrian subject, algebraic topology an American subject, algebraic geometry an Italian subject, special functions a Wisconsin subject, point-set topology a Southern subject, probability a Russian subject.

A good test for evaluating the "pop math" books that appear with clockwork regularity on the shelves of college stores is the following: let a mathematician pick one up and note whether the text is engaging enough to sustain attention for more than ten minutes. Most likely, the book will be reshelved in disgust, after the reader verifies that its contents are the usual pap of Klein bottles, chaos, and colored pictures devoid of clear meaning.

It takes an effort that is likely to go unrewarded and unappreciated to write an interesting exposition for the lay public at the cutting edge of mathematics. Most mathematicians (self-destructive and ungrateful wretches that they are, always ready to bite the hand that feeds them) turn their noses at the very thought. Little do they realize that in our science-eat-science world such expositions are the lifeline of mathematics.

When too many books are written on a subject, one of two suspicions arises: either the subject is understood and the book is easy to write—as is the case with books on real variables, convexity, projective geometry in the plane, or compact orientable surfaces. Or the subject is important, but nobody understands what is going on; such is the case with quantum field theory, the distribution of primes, pattern recognition, and cluster analysis.

Once upon a time, mathematicians would save their best papers for Festschrifts. Some of the old Festschrifts, for example, the ones in honor of H. A. Schwarz and D. Hilbert, can be read today with enjoyment; some of them have even been reprinted by Chelsea. Gradually, the quality of Festschrifts worsened, to the point that a paper appeared in *Science* magazine some twenty years ago bearing the title, "Fest Me No Schrift." Festschrifts became the dumping grounds for papers unpublishable in refereed journals.

It is bad luck to title a book "Volume One."

Masters will write masterful books. There are few exceptions to this rule. Perhaps the rule is circular, since a great master is better recognized from expository work rather than from research papers.

Richard Bellman's originality was lost because of his weakness for over-publishing, a weakness that the mathematical public does not forgive. Much as we would like to be judged by our best paper, the crowd of our fellow mathematicians will judge us by a quotient obtained by dividing the value of our best paper by the number of papers we have published.

Errett Bishop's originality was lost to the cause of constructivity. The self-appointed crowd of umpires consisting of 99% of all living mathematicians views anyone with a cause, no matter how just, as a crackpot, and from a single instance concludes that all of his papers must therefore be weird.

Both authors read better when collected in a volume: Bellman, because only his best papers are to be found (the original and clearest papers explaining his finest ideas, such as dynamic programming and invariant embedding), and Bishop, because his papers are in natural logical sequence.

Surreal numbers are an invention of the great John Conway. They will go down in history as one of the great inventions of the century. We thought Dedekind cuts had been given a decent burial, but now we realize that there is a lot more to them than meets the eye. A new theory of games lurks behind these innocent cuts. Thanks to Conway's discovery, we have a new concept of number. We will wait another fifty years before philosophers get around to telling us what surreal numbers "really" are.

A new paradise was opened when Paul Cohen invented forcing, soon to be followed by the reform of the Tarskian notion of truth, which is the idea of Boolean-valued models. Of some subjects, such as this one, one feels that an unfathomable depth of applications is at hand, which will lead to an overhaul of mathematics.

Every field of mathematics has its zenith and its nadir. The zenith of logic is model theory (we do not dare state what we believe will be its nadir). The sure sign that we are dealing with a zenith is that as we, ignorant and dumb non-logicians, attempt to read the stuff, we feel that the material should be rewritten for the benefit of a general audience.

Nowadays, if you wish to learn logic, you go to the computer science shelves at the bookstore. What lately goes by the name of logic has painted itself into a number of unpleasant corners —set theory, large cardinals, independence proofs— far removed from the logic that will endure.

When we ask for foundations of mathematics, we must first look for the unstated wishes that motivate our questions. When you search into the Western mind, you discover the craving that all things should be reduced to one, that the laws of nature should all be consequences of one law, that all principles should be reduced to one principle. It is a great Jewish idea. One God, one this, one that, one everything. We want foundations because we want oneness.

Ever since theoretical computer scientists began to upstage traditional logicians we have watched the resurgence of nonstandard logics. These new logics are feeding problems back to universal algebra, with salutary effects. Whoever believes that the theory of commutative rings is the central chapter of algebra will have to change his tune. The combination of logic and universal algebra will take over.

Combinatorial set theory deals with infinite sets and their Erdösian properties. The paradise that Cantor left us is still alive and beckons us away from the pot-boiling hell of mathematical technicians.

There is an uneasy suspicion abroad that subjects now considered distinct are one and the same when properly viewed. An example: the theory of schemes, the theory of frames, modal logic, intuitionistic logic, topoi. The unifying thread is the much-maligned theory of distributive lattices.

It has always been difficult to take quantum logic seriously. A malicious algebraist dubbed it contemptuously "poor man's von Neumann algebras." The lattice-theoretic background made people suspicious, given the bad press that lattice theory has always had.

A more accomplished example of *Desperationsphilosophie* than the philosophy of quantum mechanics is hard to conceive. It was a child born of a marriage of misunderstandings: the myth that logic has to do with Boolean algebra and the pretense that a generalization of Boolean algebra is the notion of a modular lattice. Thousands of papers confirmed to mathematicians their worst suspicions about philosophers. Such a philosophy came to an end when someone conclusively proved that those observables which are the quantum mechanical analogs of random variables cannot be described by lattice-theoretic structure alone, unlike random variables.

This *débacle* had the salutary effect of opening up the field to some honest philosophy of quantum mechanics, at the same level of honesty as the philosophy of statistics (of which we would like to see more) or the philosophy of relativity (of which we would like to see less).

We were turned off category theory by the excesses of the sixties when a loud crowd pretended to rewrite mathematics in the language of categories. Their claims have been toned down, and category theory has taken its modest place side by side with lattice theory, more pretentious than the latter, but with strong support from both Western and Eastern Masters.

One wonders why category theory has aroused such bigoted opposition. One reason may be that understanding category theory requires an awareness of analogies between disparate mathematical disciplines, and mathematicians are not interested in leaving their narrow turf.

Do you prefer lattices or categories? In the thirties, lattices were the rage: von Neumann cultivated them with passion. Then categories came along, and lattices, like poor cousins, were shoved aside. Now categories begin to show their slips. Abelian categories are here to stay; topoi are probably here to stay; triples were once here to stay, but "general" categories are probably not here to stay. Meanwhile, lattices have come back with a vengeance in combinatorics, computer science, logic and whatnot.

The term "pointless topology," goes back to von Neumann (who used to refer disparagingly to his "pointless geometries"). It has led to a serious misunderstanding. What would have happened if topologies *without* points had been discovered before topologies *with* points, or if Grothendieck had known the theory of distributive lattices?

We thought that the generalizations of the notion of space had ended with *topoi*, but we were mistaken. We probably know less about space now than we pretended to know fifty years ago. As mathematics progresses, our understanding of it regresses.

Spaces that are disconnected in any one of many ways have become the last refuge for the concept of a topological space that the Founding Fathers had meant.

Universal algebra has made it. Not, as the founders wanted, as the unified language for algebra, but rather, as the proper language for the unforeseen and fascinating algebraic systems that are being discovered in computer algebra, like the fauna of a new continent.

The concept of a partially ordered set is fundamental to combinatorics. What seems like a miracle is how theorems on partially ordered sets that were discovered long ago without combinatorial motivation have turned out to be the key to the solution of combinatorial problems. But then, why do such coincidences surprise us? Isn't this surprise what mathematics is about?

Graph theory, like lattice theory, is the whipping boy of mathematicians in need of concealing their feelings of insecurity.

The theory of invariants is related to the theory of group representations in much the same way as probability theory is related to measure theory. A functional analyst could spend a lifetime with measurable functions without ever suspecting the existence of the normal distribution. Similarly, an algebraist could spend a lifetime constructing representations of his favorite group without ever suspecting the existence of perpetuants.

The parallel can be carried further. The justification of probabilistic reasoning by measure theory has never been fully convincing. For one thing, there seems to be no hope of proving that probability leads categorically to measure theory; noncommutative and nonstandard probabilities keep reemerging every few years. More important, probabilistic reasoning is syntactic, whereas its measure theoretic justification is highly semantic. Probability remains a rare instance of a syntax that has been justified only by the mercy of one known semantic model.

The case of invariant theory is worse. Here the syntactical language of invariant theoretic reasoning, commonly spoken one hundred years ago in Cambridge (U.K.), Naples, Erlangen, or Paris, has died without a trace. Only the writings remain, a pale mirror of what at one time must have been the living spirit of algebra. What we have instead is the theory of group representations, now rich enough to be called by several names (Lie theory, algebraic groups, symmetric spaces, harmonic analysis, etc.) which, as the subspecialists in these subjects will vociferously insist, are supposed to stand for separate and independent fields.

It is a task for the present generation to recreate the lost life of invariant theory. Whether this task will be accomplished by rereading and reinterpreting the classics or whether it will be reinvented, as some physicists are now doing, will be one of the dramas of the coming years which will be watched with interest by the few who care.

Meanwhile, it may not be remiss of those mathematicians who still believe in history to reissue and eventually modernize nineteenth-century invariant theory, unquestionably the greatest untapped inheritance from our great-grandfathers.

After years of the dictatorship of groups, semigroups are asserting their rights, and after some affirmative action we may get courses on semigroups taught at Harvard and Princeton.

One remarkable fact of applied mathematics is the ubiquitous appearance of divergent series, hypocritically renamed asymptotic expansions. Isn't it a scandal that we teach convergent series to our sophomores and do not tell them that few, if any, of the series they will meet will converge? The challenge of explaining what an asymptotic expansion is ranks among the outstanding but taboo problems of mathematics.

Lebesgue said that every mathematician should be something of a naturalist.

Some chapters of mathematics have been pronounced dead several times over. But like the Arabian phoenix they rightfully claim, "*eadem resurgo*." The last time the theory of summability of divergent series was declared obsolete was after the publication of Laurent Schwartz's theory of distributions. This theory was supposed to do away with a lot of things: summability, Banach spaces, functions. Instead, the theory of distributions, after acquiring some degree of acceptance among specialists in the more absurd reaches of partial differential equations, has dropped into oblivion (while Banach spaces are flourishing more than ever). Sometime in the fifties Alberto Calderón remarked: "As soon as people realize that they cannot make changes of variables, the theory of distributions will be in trouble."

Every graduate student should study inequalities for at least one term. They are likely to be useful for life, no matter what field of mathematics is later chosen. Finding a theory for some lovely inequality is a research challenge for mathematicians with limited backgrounds. Most inequalities do not fit anywhere; others are crying for theorylets of their own.

Fascination with inequalities will never end; the Landau inequality relating the norms of a function, its first derivative and its second derivative is equivalent to Heisenberg's uncertainty principle.

Hilbert space seemed well run through a few years ago (how pathological can an operator be?) but the theory of linear operators has found two new lives: by remarriage with circuit theory and by a love affair with systems theory.

The late William Feller was fond of repeating: "I could stare all my life at a symmetric matrix, I will never get Hilbert space out of it!"

Some ideas of mathematical analysis have been independently discovered in circuit theory. This cannot be a coincidence (there are no coincidences in mathematics), but no one seems to have investigated what lies behind this strange parallelism, rewarding as such an investigation might be.

Among all subjects in the undergraduate mathematics curriculum, partial differential equations is the hardest to teach. Disparate subjects are brought together by the sole fact of having $\frac{\partial}{\partial x}$ in common—making strange bedfellows.

It used to be called the calculus of variations and everybody thought it was as dull as night. Now it is called optimization theory, and everybody thinks it is red hot.

The soliton fad is a catchy mixture of operator algebra and explicitly solvable differential equations. How can anyone resist such a temptation? So much for the good news. The bad news is that solitons seem to be unbudgingly one-dimensional, despite the insinuations of computer simulators. Chalk one up to one-dimensional physics.

When a publishing company is about to go out of business after having published too many advanced (and hence unsaleable) mathematics books, the managers are well advised to recover their losses by muscling into the field of statistics. From a marketing point of view, statistics books are in a class with cookbooks and mystery books: they have an assured market, and the intellectual demands on the reader are modest. A fellow with a $40 monthly book allowance will be tempted to spend it on a statistics book. The reading is likely to be easy, the material will have some applications, and it may suggest topics for research.

As the fortunes of pure mathematics decline, the sorts of statistics rise. Anyone who wants to make a fast buck after an undergraduate education in mathematics (90% of our majors) has a choice between statistics and computer science. It looked for a while like computer science was about to wipe out statistics, but the tables are turning.

Books on paradoxes in statistics are similar to mystery books. They have a faithful readership, and they follow a rigorous sequence in their presentation, like Greek tragedies.

It may appear surprising that there are so few such books, considering the readership, but the possible tables of contents for books on paradoxes are severely limited. It is hard to come up with a new paradox, even by yielding to the sleaziest equivocations of statistical "reasoning." A mix of old paradoxes, together with some clever remarks, makes good bedside reading, though not quite as good as Raymond Chandler.

One cannot overcome the suspicion that Lebesgue's star is fading faster than that of other French mathematicians, such as Paul Lévy, who could not have cared less about the Lebesgue integral.

In the world of probability, good exposition is a tradition that not only keeps the field alive but makes the results of the latest research available to scientists who depend on probability for their livelihood. How we wish we could say the same for topology! (When will topologists wake up to the harsh realities of the nineties?)

Writers of probability texts are laboring under the illusion nurtured by the late Willy Feller that discrete probability is easier than continuous probability.

The idea of proving the existence of a mathematical object by proving that the probability of its existence is positive is one of the most fertile ideas to come out of the great Erdös. There is a large number of objects whose existence can only be proven probabilistically, and for which no construction is known.

Professional logicians should become interested in this method of proving existence theorems. There ought to be a correspondence principle allowing us to translate probabilistic algorithms into combinatorial algorithms or, lacking such a translation, a new logic associated with probabilistic reasoning. Unfortunately, those people who are competent to carry out such a program have better fish to fry, and philosophers of mathematics are too incompetent to deal with a problem which ought to be their bailiwick.

Two problems of probability dismissed as old-fashioned keep resurfacing with predictable frequency. They are: (1) the relation between a multivariate probability distribution and its marginals, and (2) inequalities holding among the cumulants of a probability distribution. Neither of them is satisfactorily solved; like thorns in our probabilistic flesh, they are reminders that we have not been able to go one up on our nineteenth-century ancestors. The author remembers the late Willy Feller remarking, "It is a scandal that we should not be able to handle cumulants in this day and age!"

What is the difference between a textbook on probability and one on probability modeling? More generally, what is the difference between a textbook on X and one on X-modeling? Let me tell you. A textbook on probability is expected to carry certain subjects, not because these subjects are useful or educational but because teachers expect them to be there. If such a textbook omits one of the liturgic chapters, then Professor Neanderthal will not adopt it, and the textbook will go out of print.

As a consequence of this situation, most elementary probability textbooks are carbon copies of each other. The only variations are colors and settings of the exercises. The only way to add a topic to the standard course is to name it "probabilistic modeling." In this devious way, Professor Neanderthal may be persuaded to adopt the additional text as "supplementary reading material." The students will find it interesting, and eventually the topic will be required. What is missing is someone willing to (1) set his research work aside to write an elementary modeling book (thereby risking permanent status of *déclassé*), (2) take the trouble to rethink the subject at an expository level (a harder task than is generally assumed), and (3) find a publisher who will take a chance on the new text (increasingly difficult).

The estimation of rates of convergence in the central limit theorem keeps the central limit theorem from being a psychological theorem. It used to be a Russian industry (state-owned of course), but more recently the capitalist West has begun to find it profitable, though the new wave has not yet reached Singapore, South Korea, or Taiwan.

The distribution of the maxima of n independent — or dependent — random variables is a problem that fifty years ago might have been dismissed as worthy of a brilliant note, or a footnote by Paul Lévy. Instead, it has blossomed into a fascinating chapter of probability theory, a testing ground for the most ingenious combinatorial techniques of our day.

Conditional probability tends to be viewed as one technique for calculating probabilities. Actually, there is more to it than meets the eye: it has an interpretation which is ordinarily passed over in silence. Philosophers take notice.

Stochastic geometry is a growing offspring of the marriage between integral geometry and the theory of point stochastic processes. It exploits an old idea of algebraic geometry: when studying a family of elements subject to reasonable conditions, make a variety out of them and study the conditions as subvarieties.

Unbeknownst to the world, a new school of probabilists, under the leadership of the great Kesten has been hatching the stochastic processes of the future and breaking the one-dimensional barrier, which in more Doobian times seemed impenetrable. The combinatorics of percolation, the anfractuosities of many-particle motion, and the mysteries of phase transition are at last within the reach of mathematical rigor. We hope never to be asked to teach the course: it might take many years of study.

Percolation is one of the most fascinating chapters of probability. The difficulty of the subject is not commensurate with its naturalness; one expects something so beautiful to have a pretty proof here and there, and not to be engulfed by heuristic arguments.

Probability has crashed the dimension barrier. A few years ago self-styled Cassandras were proclaiming the death of probability on the Procrustean bed of the real line. Markov processes, and even, mistakenly, martingales, were viewed as the rightful heirs of linear stochasticity. As soon as Markov random fields showed the way to go many-dimensional and keep Markovianness to boot, these bad dreams ended. Things started to pop when point processes came along. If there ever was an open field with wide frontiers, this is it. Do not listen to the prophets of doom who preach that every point process will eventually be found out to be a Poisson process in disguise.

The distance between probability and statistical mechanics is diminishing, and soon we won't be able to tell which is which. We will be rid of the handwaving arguments with which mathematically illiterate physicists have been pestering us.

Statistical mechanics is the first bastion of sound physics in this century. Physicists have discovered the usefulness of good mathematics and rigorous theorems. In other branches of physics, physicists make a curious distinction between "proofs" and "rigorous proofs," falling prey to a *folie* of mystical trance in which they can make themselves believe anything.

Monographs in queueing theory appear with the regularity of a Poisson process of high intensity. One wonders what motivates such prolific writing, especially as the term "queueing theory" does not awe us like secondary obstructions and higher reciprocity laws. A malicious observer will infer that it is easy to write books on queueing theory. When the monograph is written by a Russian, as several were every year until a few years ago, one may explain the phenomenon by appealing to Russia's tradition in probability. But when even a Frenchman writes one, then the field is enjoying a fashion of sorts. Queueing theory is that rarest branch of mathematics: it uses (and even helps develop as it did with Hille) high-grade functional analysis: the theory of semigroups of operators and spectral theory. It deals with stochastic processes without getting enmeshed in measurability meshes; it provides computable yet useful examples which bring into play fancy special functions, and it relates to practical reality.

Of some subjects we may say they exist, of others, that we wish they existed. Cluster analysis is one of the latter.

Pattern recognition is big business today. Too bad that none of the self-styled specialists in the subject — let us charitably admit it *is* a subject — know mathematics, even those who know how to read and write.

It is a standing bet between physicists and mathematicians that thermodynamics cannot be axiomatized.

The maximum entropy principle is one of the hot potatoes of our day. It has not split the world of statistics as has Bayes' law, but no one has yet succeeded in finding a justification for it. Maybe we should make it one of the axioms of statistics.

The Feynman path integral is the mathematicians' *pons asinorum*. Attempts to put it on a sound footing have generated more mathematics than any subject in physics since the hydrogen atom. To no avail. The mystery remains, and it will stay with us for a long time.

The Feynman integral, one of the most useful ideas in physics, stands as a challenge to mathematicians. While formally similar to Brownian motion, and while admitting some of the same manipulations as the ones that were made rigorous long ago for Brownian motion, it has withstood all attempts at rigor. Behind the Feynman integral there lurks an even more enticing (and even less rigorous) concept: that of an amplitude which is meant to be the quantum-mechanical analog of probability (one gets probabilities by taking the absolute values of amplitudes and squaring them: hence the slogan "quantum mechanics is the imaginary square root of probability theory"). A concept similar to that of a sample space should be brought into existence for amplitudes and quantum mechanics should be developed starting from this concept.

The theory of functional integration is one of those rare meeting points of several branches of mathematics. There is the algebra of creation and annihilation operators, the topology of Feynman diagrams, the combinatorics of Wick products, the analysis of the normal distribution, the probability of Wiener integrals, the physics of second quantization.

Ergodic theory is torn between two extremes: on the one hand, it pretends to be a generalization of the mechanics of flows (well, measurable flows instead of continuous flows, but let it pass); on the other hand, it is the study of shift operators originating from stationary stochastic processes and their generalizations.

Slowly but inevitably, the basic concepts of ergodic theory are being expanded to include thermodynamics. The latest newcomer into the rigorous fold is pressure. Maybe the long-awaited axiomatization of thermodynamics is within reach.

The tendency towards concreteness is most evident in the funneling of ergodic theory into what used to be called "phase plane analysis" and is now rebaptized as "propagation of chaos." The remarriage between Julia-style functional iteration and the dynamical theory in the plane is bearing an unexpectedly good-looking offspring.

Some old notions from dynamical systems theory, going back to Poincaré and G. D. Birkhoff, have been brushed off the shelves and resold to an unsuspecting public as the *dernier cri*. Such a selling technique is not rare in mathematics (the author himself admits to having rewritten some of his own twenty-year-old papers in a new notation in a vain attempt to get someone to read them).

The ecstatic words "chaos" and "fractals" have revived the study of dynamical systems. They prove the power of new terminology injected into a staid theory.

When computers came along and everyone was talking about the end of mathematics, it was thought that special functions would be one of the first casualties. The opposite has happened: special functions are going great guns as never before, not even in Kummer's time. It is hard to tell who will win the fraternal joust between the abstract but penetrating concepts of the theory of group representations and the concrete but permanent formulas of the theory of special functions. We place our bets on the latter.

Hypergeometric functions are one of the paradises of nineteenth-century mathematics that remain unknown to mathematicians of our day. Hypergeometric functions of several variables are an even better paradise: they will soon crop up in just about everything.

Dilogarithms would not have been given a chance fifty years ago, not even by someone who contributed to the Salvation Army. Along came I. M. Gel'fand and related them to Grassmannians, and people started to take notice. Now they are about to break into High Society, and the bets are on that they will be seen at the next debutantes' ball.

Yesterday it was differential equations, today it is group representations, tomorrow it may be second quantization. Behind these fields there is one invariant: the theory of special functions in its nudity and a list of identities.

Suddenly, resultants are "in" after a lapse of one hundred years, during which André Weil's preposterous *"Il faut éliminer l'élimination"* was taken seriously. We have a marvelous new area of research for the algebraist, the algebraic geometer, and most of all the combinatorialist. We are back to year zero after one hundred years of abstract algebraic geometry.

After fifty years of commutative algebra, we can go back to thinking geometrically without the unmentionable fears that haunted our fathers.

Every field has its taboos. In algebraic geometry the taboos are (1) writing a draft that can be followed by anyone but two or three of one's closest friends, (2) claiming that a result has applications, (3) mentioning the word "combinatorial," and (4) claiming that algebraic geometry existed before Grothendieck (only some handwaving references to "the Italians" are allowed provided they are not supported by specific references).

Algebraic geometry is coming of age, and introductions are getting better. They are never easy to read. The subject is difficult to motivate, but it is so venerable that one feels inclined to engage in the nontrivial effort of reading such introductions, despite the gnawing suspicion that one may be wasting one's time.

Textbooks in algebraic geometry should be written by Italians and corrected by Germans.

For a long time, books on algebraic topology were of two kinds: either they ended with the Klein bottle, or they were written in the style of a letter to Norman Steenrod.

In number theory, clearly written books are a rarity: the average book requires a modest background of two years of algebraic geometry, two years of analytic number theory, a working knowledge of class field theory and p-adic analysis. How the mathematics publishing industry can survive publishing books in number theory is a mystery which we will charitably refrain from investigating.

The neglect of exterior algebra is the mathematical tragedy of this century. Only now is it slowly being corrected. Who cannot marvel at that joint success of combinatorics, algebra, and analysis that is the theory of exterior differential systems?

These systems were introduced to describe trajectories of motion. Yet, the intrusion of geometric notions is like the appearance at a wedding of a close but ill-dressed relative. The algebra should dictate the semantics, as has happened in algebraic geometry, logic, etc. Whatever spatial notions will be associated to exterior differential systems should be obtained by an unprejudiced model theory inexorably determined by the syntax of exterior differential algebra.

No proper textbook presentation of exterior algebra has been written. Both van der Waerden (a student of General Weitzenböck) and Emmy Noether (a student of Gordan) hated classical invariant theory and everything that went with it. In their first textbook of "modern" algebra, they made sure that not a word of tensor algebra, and most of all, no hint of mystifying concoctions of that crackpot Grassmann would be given out. Emil Artin did not learn exterior algebra until late in his career. The first exposition of exterior algebra came with Bourbaki's *Algèbre*, Chapter Three (the first edition is better than the second); but Bourbaki did not believe in geometry and failed to provide the translation of geometry into algebra that exterior algebra makes possible. A treatise in this fundamental chapter of mathematics is awaiting its author; meanwhile, we have to bear with mathematicians who are exterior algebra-blind.

Supersymmetric algebra sheds new light on the notion of space. We will have two ways of doing geometry and physics: old-fashioned and supersymmetric. It was once thought that supersymmetric analogs of the facts of physics and geometry were "purely formal," without a "real" interpretation. The opposite is true: supersymmetric algebra expresses in a language that is as clear as it is simple some of the deepest facts about space and time. With the advent of supersymmetric algebra, tensor algebra has found a lucid notation which is likely to be definitive.

The amount of time lost in rediscovering the elementary conclusions of phenomenology is matched by the arrogance with which latter-day self-styled philosophers of perception promulgate their rediscoveries. We are lacking an adequate translation of Husserl's texts. Husserl is no easier to translate into English than a Japanese haiku, and it is not realistic to expect an A.I. technician to learn German. Something must be done before the know-nothings take over.

Over the years, the constant and most reliable support of science has been the defense establishment. While old men in congresses and parliaments debated the allocation of a few thousand dollars, farsighted generals and admirals would not hesitate to divert substantial sums to help the oddballs in Princeton, Cambridge, and Los Alamos. Ever since Einstein wrote that letter to President Roosevelt, our best friends have been in that branch of government concerned with defense. And now that intelligence collection is reaching byzantine complexity, we can learn to appreciate another source of support that may be coming along.

"Historia magistra vitae, lux veritatis. . . ."

Reading the classics is more important for the student than the perusal of certain journals pretending to publish so-called advances in mathematics.

Leonardo da Vinci wrote: "Theory is the captain and application is the soldier."

CHAPTER XXI

Book Reviews

Paul Halmos: a Life

Every mathematician will rank other mathematicians linearly according to past accomplishments while he rates himself on the promise of his future publications. Unlike most mathematicians Halmos has taken the unusual step of printing the results of his lifelong ratings. He is pretty fair to everyone included in his lists (from first-rate [Hilbert] to fifth-rate [almost everybody else]), except to himself, to whom he is merciless (even in the choice of the title to the book:[1] "I *Want* to be a Mathematician," as if there were any question in anybody's mind as to his professional qualifications).

Those scientists who take the uncustomary step of writing their autobiography are often driven by a desire to suppress that "drop-in-the-bucket" feeling about their work. This disease strikes mostly in old age, an undeserved punishment at the end of many a productive career. In the majority of cases, autobiography has not proven to be good therapy. What is worse, the attempted cure will result in a pathetic "see-how-good-I-am" yarn of achievements and honors, a motley assortment of ephemeral episodes that will be soon forgotten if read at all.

Halmos has discovered another therapy, one that may work. He admits from the start his own limitations (he overadmits them). Once he has taken this courageous first step, the task of telling the truth, which for the average autobiographer is a formidable moral imperative, becomes for him an easy exercise in dispassionate description. And describe he does, down to the minutest detail: the colors of the cups in the tearoom at the University of Chicago, the number of pages in the bluebooks at the University of Hawaii, the bad quality of the chalk at Moscow S.U. The leading thread of his exposition, making his narration entertaining and not just interesting, is mathematical gossip, which is freely allowed to unfold in accordance with its mysterious logic. The reader will be thankful for being spared the nauseating personal details that make most autobiographies painful reading experiences ("My family was very musical," "The winter of 1932 was exceptionally severe"). Whatever does not relate to the world of mathematics is ruthlessly omitted (we hardly even learn whether he has a wife and kids).

At last we have a thorough account (one that stands the test of rereading, and the only one of such kind) of the period that runs approximately from the forties to the present day, a period that will go down in history as one of the golden ages of mathematics. However, the theme that emerges from this collection of amusing anecdotes is not the welcome lesson we would expect as the bequest of a golden age. Halmos' tales of incompetent department heads, Neanderthal deans, and obnoxious graduate students reveal, in the glaring light of gossip, the constant bungling, lack of common sense, absence of *savoir faire* that is endemic in mathematics departments everywhere. Take, for example, the turning point of the author's career, his departure from the University of Chicago. Even granting Halmos' contention that his papers may have lacked depth (at least in someone's opinion) in comparison with those of certain colleagues of his, it still seems clear that the university made a mistake by dispensing with Halmos' services. Whatever his other merits, Halmos is now regarded as the best expositor of mathematics of his time. His textbooks have had an immense influence on the development of mathematics since the fifties, especially because of their influence on young budding mathematicians. Halmos' glamor would have been a sounder asset to the University of Chicago than the deep but dull results of an array of skillful artisans. What triumphed at the time was an idea that still holds sway in mathematics departments, namely the simplistic view of mathematics as a linear progression of problems solved and theorems proved, in which any other function that may contribute to the well-being of the field (most significantly that of exposition) is to be valued roughly on a par with the work of a janitor. It is as if in the filming of a movie all credits were to be granted to the scriptwriter at the expense of other contributors (actors, directors, costume designers, musicians) whose roles are equally essential for the movie's success.

This well-worn tirade would hardly be worth repeating (and the author manages to keep it *sub rosa*) were it not for the fact that the mathematicians' share of the scientific pie is now shrinking. A strong case can be made that mathematics is today the healthiest and (what now begins to matter) the most honest of the sciences. But just knowing the truth is not enough: the outside world must be made to believe it (it takes two to tango). Whereas physicists, chemists, and biologists have learned to properly appreciate everyone and to justly apportion rewards to all (much like credits in a movie), even today (and even after reading Halmos), too many mathematicians, when confronted with the painful chore of protecting (defending, asserting, popularizing, selling) mathematics would rather be like the captain of a sinking ship, bravely saluting while going down, the Clifton Webb in the sinking of the *Titanic*.

The Leading Line of Schaum's Outlines

When you travel to Naples, Zimbabwe, Bombay, or Dresden and visit the scientific bookstores of such blessed places, you will probably not find on their shelves any volume of Springer's *Grundlehren* series, as may have been the case in the golden times of Courant and Hilbert, or any of the numerous textbooks by Serge Lang, or a volume of Springer's flagship series, *Graduate Texts in Mathematics*, or even any of the endemic publications of the American Mathematical Society. In Togo and Guyana these publications are unknown, and the math departments of the most distinguished universities in these countries are not on the mailing list of the Springer Newsletter.

In place of such illustrious book series, every scientific bookstore from Santiago to Islamabad, from Nigeria to Indonesia to Ecuador to Greece, makes a point of shelving the nicely bound, elegantly printed volumes known in this country as *The Schaum's Outlines*.[2] A wide selection of titles from this series is likely to be the backbone of scientific bookstore sales. Individual volumes may be purchased at a price affordable to college students, either in the original English or in bilingual editions—indispensable for ambitious students eager to pursue their graduate studies in the U.S. with an eye towards an eventual green card— brilliantly translated by mathematicians of high local renown into Urdu, Papiamento, Cantonese, Basque, even Esperanto.

Why is it that these much-maligned (but far from neglected) presentations of the fundamentals of mathematics— from calculus to differential geometry, statistics to tensor algebra, thermodynamics to quantum field theory—are profitably sold in every scientific bookstore while the acclaimed classic texts in the same areas, those standard references of the happy few in Cambridge, New Haven, Princeton, Stanford and Berkeley, have trouble selling a few hundred copies worldwide? Why is it that Serge Lang's *Linear Algebra*, published by no less a Verlag than Springer, ostentatiously displays the sale of a few thousand copies over a period of fifteen years, while the same title by Seymour Lipschutz in *The Schaum's Outlines* will be considered a failure unless it brings in a steady annual income from the sale of a few hundred thousand copies in twenty-six languages?

Mathematicians, born snobs that they are, have a ready answer to explain away this embarrassing defalcation of royalties. It is an answer that does not speak its name, lest one be labelled "politically incorrect." Mathematicians would like to put themselves to sleep by a convenient bromide. They would have everyone believe that the texts published under the banner of *The Schaum's Outlines* are inferior products, suitable for inferior countries, inferior races, inferior universities, and inferior teachers— the inevitable outcome of mass culture, like children's toys from Taiwan, rock music, Hollywood movies, and television's *Dallas*.

Such clichés may for a while succeed in keeping everyone at a safe distance from the hard realities of life. This facile explanation is not only wrong but it lies at the heart of the disastrous state of the teaching of mathematics, which is threatening to drag our profession into an irreversible collapse. A proper understanding and appreciation of the success of the *The Schaum's Outlines* will be a cleansing experience, designed to rid ourselves of wishful thinking, parochial explanations, and unwarranted one-upmanship.

This writer proudly acknowledges his kinship and wholehearted solidarity with those "inferior" people who peruse the volumes of *The Schaum's Outlines*. On several occasions, he has found in these outlines a ready, concise, clear explanation of some point in mathematics.

I vividly recall one case. I had to know what the analysis of variance was, for reasons better forgotten. I was in a hurry, probably because I had to answer a query by a non-mathematician. In vain did I peruse various standard textbooks in statistics, only to discover the weaknesses in their presentations, and to feel irritated by their clumsy expositions. To wit:

(a) In describing the analysis of variance which appears in Chapter Seven, tacit reference is made to various unspecified previous chapters, all of which the reader is supposed to have mastered without notice.

(b) Some of the notation used in Chapter Seven is not explained anywhere. One has to leaf through the preceding 300 pages to uncover the meaning of even some of the symbols which the author deigns to explain. Nor is an index of notation to be found anywhere. Some symbols are given different meanings in different chapters; for other symbols, the meaning is taken for granted— as in a discussion among friends.

(c) A dangerous confusion is made between description and definition.

The author believes that the reader can master the analysis of variance by reading its definition. No statement is made as to what motivates the analysis of variance, why it is important, who uses it and why people should pay money to learn it. Examples are either trivial or come much later on, and then they take up entire chapters. No "typical example" is given.

(d) The author seems to believe the Landau style should be the ideal of mathematical exposition. The destructive influence on mathematical exposition of Landau's notorious tracts is far from over. A sizable portion of the mathematical community still believes in the educational value of Landau's cryptography. No concession is made to the reader, who at every step is expected to decipher rather than read, to puzzle over rather than understand, to be baffled rather than enjoy the reading. After consulting several such textbooks, each one guilty of one or more of the above, I turned in desperation to Murray Spiegel's *Statistics*, a bestselling volume in the *The Schaum's Outlines*, now in its fifth printing of a second edition. I read with interest the following significant example:

> A company wishes to purchase one of five different machines: A, B, C, D, or E. In an experiment designed to test whether there is a difference in the machines' performance, each of five experienced operators works on each of the machines for equal times. Table 126.11 shows the numbers of units produced per machine. Test the hypothesis that there is no difference between the machines at significance levels (a) 0.05 and (b) 0.01.

After reading this paragraph, I knew what the analysis of variance is. I also realized why students of statistics in France, Italy, and other countries on the continent systematically shun the pompous and incomprehensible texts of the local professoriate in favor of Murray Spiegel's book. I maintain that Murray Spiegel's *Statistics* is the clearest introduction to statistics in print.

I can almost hear the reader's next bromide: "Well, these are texts written for *applied* mathematicians!"

Half the volumes in the series are counterexamples. Elliott Mendelson's *Boolean Algebra*, for example, is no more applied than the three-volume survey published by North-Holland (which is equally fascinating). But what Elliott Mendelson realizes, and what most textbooks withhold, is that one of the best ways, possibly *the* best way, of understanding Boolean algebra is through visualization by switching circuits. Important discoveries in Boolean algebra were made by mathematicians (such as Claude Shannon in his celebrated master's thesis) who visualized "and" and "or" by switching circuits. Quine's theory of prime implicants, strangely neglected by the standard texts (perhaps because it is the most useful chapter of the theory), becomes obvious in the circuit-theoretic interpretation. The numerous axiomatizations of Boolean algebra (no other branch of mathematics has received such a bounty of axiomatizations) turn into a series of intriguing puzzles when the reader attempts to visualize them by circuits.

All basic notions of Boolean algebra are given concise, definitive presentations by Mendelson. Free Boolean algebras are presented in a simpler fashion than any other published author (most authors feel themselves obliged to digress into universal algebra at this point, for liturgical reasons). Elliott Mendelson's *Schaum's Outline on Boolean Algebra* is the best introduction to Boolean algebra ever written.

What lies behind the success of *Schaum's Outlines*? Why should we consider Spiegel, Mendelson and other authors of these outstanding expositions benefactors of mankind?

The best introduction to mathematics is not achieved by rigorous presentation. No one can learn calculus, linear algebra, or group theory by reading an axiomatic presentation. What one wishes is a feeling for a piece of mathematics. Let the student work with unrigorous concepts that lead as quickly as possible to a half-baked understanding of the main results and their applications. Rigorous presentation can occur later, as an afterthought, to be given only to a fraction of the students, those who develop a genuine interest in mathematics.

It is a dreadful mistake to expect that everyone who learns an area of mathematics should be subjected to the learning of the foundations of that particular area. This mistake was made in the sixties, with the introduction of new math, a step backwards in the teaching of mathematics from which we are a long way from recovering. The student needs to develop an understanding, however partial and imperfect, by descriptions rather than definitions, by typical examples rather than grandiose theorems, by working out dozens of menial exercises.

Most mathematicians who teach mathematics fail. They bask in the illusion that the majority of their students should become mathematicians, or their teaching is wasted; or in the illusion of immediate, effortless understanding, the illusion that it suffices to present the "facts" and let students figure out the "sense" of the mathematics, a sense that substantially differs from the "meaning." Authors of *Schaum's Outlines* have learned to avoid these illusions. Anyone who is about to teach the undergraduate mathematics curriculum should come down to earth by looking through *The Schaum's Outlines* before burdening the class with those well printed, many-colored, highly advertised hardcover volumes that are pathetically passed off as textbooks.

Professor Neanderthal's World

Every field of mathematics has a lifespan of its own. Take noncommutative ring theory. Starting with a period of giddy exuberance (Wedderburn, Dickson), the subject then enjoyed a lengthy and dignified maturity (Jacobson, Albert) when the basic problems were either solved or properly reformulated. Then came old age. A heap of leftover problems is forced to yield to the accrued power of a technique achieved by generations of mathematicians. Finally, on the day of *redde rationem*, there may or may not come along some savior, in this case T. Y. Lam.[3] Lam gives the definitive presentation, so that everyone in the mathematics community has the opportunity to become aware of whatever noncommutative ring theory has to offer.

Few areas of mathematics have found their T. Y. Lam and most of them have simply withered. Retrieving results, even fundamental ones, is a painful search that in another area of learning would be carried out by professional historians. Most historians of mathematics, however, hardly know how to differentiate and integrate, and they lack the mathematical *Kultur* for an updated presentation of the mathematics of yesteryear.

Thus, there is the danger that every new generation of mathematicians will rediscover the results of a hundred years ago, to everyone's dismay. Worse yet, the main customers of mathematics are physicists and engineers, whether we like it or not, and they have given up searching for relevant results in unreadable research articles. They will rediscover by themselves whatever mathematics they need, hoping to shove mathematicians into the dustbin of history.

We will set aside an examination of the tragic consequences of this catastrophic situation. The reasons for the lack of a proper exposition of most areas of mathematics lie deep in our mathematical psyches. For fear of reprisals, we shall put forward only two.

1. By and large mathematicians write for the exclusive benefit of other mathematicians in their own field even when they lapse into "expository" work. A specialist in quantum groups will write only for the benefit and approval of other specialists in quantum groups. A leader in the theory of pseudo-parabolic partial differential equations in quasi-convex domains will not stoop to being understood by specialists in quasi-parabolic partial differential equations in pseudo-convex domains. Such "expositions" are more often than not brilliant displays of virtuosity, designed to show the rest of the community (a half-dozen individuals) how much more elegantly and simply the author would have proved somebody else's results were it not for his more important commitments.

2. The bane of expository work is Professor Neanderthal of Redwood Poly. In his time, Professor Neanderthal studied noncommutative ring theory with the great X, and over the years, despite heavy teaching and administrative commitments (including a deanship), he has found time to publish three notes on idempotents (of which he is justly proud) in the *Proceedings of the American Mathematical Society*.

Professor Neanderthal has not failed to keep up with the latest developments in noncommutative ring theory. Given more time, he would surely have written the definitive treatment of the subject. After buying his copy of T. Y. Lam's long-expected treatise at his local college bookstore, Professor Neanderthal will spend a few days perusing the volume, after which he will be confirmed in his darkest suspicions: the author does not include even a mention, let alone a proof, of the Worpitzky-Yamamoto theorem! Never mind that the Worpitzky-Yamamoto theorem is an isolated result known only to a few initiates (or perverts, as graduate students whisper behind the professor's back). In Professor Neanderthal's head this omission of his favorite result is serious enough to damn the whole work. It matters little that all the main facts on noncommutative rings are given the clearest exposition ever, with definitive proofs, the right examples, and a well thought out logical sequence respecting the history of the subject. By judicious use of the U.S. mails, the telephone, and his recently acquired fax machine, Professor Neanderthal will make sure that the entire noncommutative ring community will be made aware of this shocking omission. He will send an unsolicited review of the book to the *Bulletin of the American Mathematical Society* in which the author's *gaffe* will be publicized.

At least half of the sales of *A First Course in Noncommutative Rings* (excluding library sales) are accounted for by the Professor Neanderthals of this world. How many potential authors of definitive treatises have the patience, stamina, and spirit of sacrifice that will allow them to withstand the onslaughts of snide criticisms, envious remarks, uncalled-for categorical assertions from the less endowed members of their community? How many mathematicians are willing to risk the snubs of their peers for the welfare of mathematics? T. Y. Lam does, and we thank him for it. Not one iota would we like to see changed in this book, we poor helots once shown the door with a pointed finger by members of the noncommutative ring community. We have enjoyed reading T. Y. Lam in bed, and his book will take its place together with eleven other classics of mathematics as one of our *livres de chevet*.

Uses and Misuses of Numbers

In this otherwise laudable book,[4] exoteric and esoteric math are dangerously mixed and presented as if they were one and the same. We believe the editors have made a serious mistake.

Most of the topics are well-known to the average mathematician; the would-be mathematician would not stoop to learning from such a book. Potential buyers of the text are students, amateurs, and mathematicians who delight in rereading what they already know. This last category makes up a non-negligible portion of book buyers.

Among the exoteric subjects are the all-too-short chapter on p-adic numbers, the one on algebras with composition laws, the one on Hurwitz' theorem, the extremely interesting exposition of division rings, the minicourse on non-standard analysis, and the beautifully written chapter on numbers and games. From each of them, the lay reader will learn some new mathematical facts with unexpected applications.

The same cannot be said of the remaining chapters, dealing with natural numbers, integers and rational numbers, with real and complex numbers, the fundamental theorem of algebra, π, the isomorphism theorems of Frobenius, Hopf, and Gel'fand and Mazur, and set theory. There is nothing more deadly for a non-professional mathematician than being assaulted by a rigorous version of facts which have always been obvious. The presentation of the real numbers by Dedekind cuts, for example (Chapter 2), or the description of the integers starting with the Egyptians and the Babylonians (Chapter 1) may be fascinating subjects to those who are aware of the plethora of different algebraic number fields which one learns after studying math for six straight years, but they are a sure turnoff for students in engineering, English, or Polynesian dances who have to take one course to fulfill their math requirement. These poor students would like to be told something they did not already know or suspect.

The chapter on π is particularly obnoxious. It contains a list of pointless characterizations of that wonderful number, as well as a host of formulas adding up to π. No amount of classroom dramatization will succeed in making this rattling off of unmotivated formulas palatable.

The two chapters on real and complex numbers are downright dangerous. Giving a rigorous presentation of the obvious has always been the dark side of mathematics, the side that makes pure math disliked and unwelcome, and threatens periodically to eliminate its being taught altogether.

On the exoteric side, however, the chapters on division algebras and topology tell every reader that mathematics has something new and fascinating to contribute, something which the reader has never fathomed. The same can be said of the chapter on numbers and games, where — would you believe it? — the dreaded Dedekind cuts make their unexpected reappearance as the springboard for the most beautiful generalization of the concept of number since Cantor, which is the great discovery of the great Conway. Similarly, one reads with interest about the possibilities of non-standard analysis.

A few chapters, perched between the exoteric and the esoteric, should be considered obsolete: on Hamilton's quaternions and Cayley numbers or alternative division algebras. It would have been best to omit them entirely, along with the chapter on complex numbers. Alternatively, a substantial chapter on Clifford algebras would have been welcome. A light description of the field of complex numbers could be shrewdly followed by an unexpected discussion of the analogy between complex numbers and Clifford algebras, showing how Cauchy's integral formula and linear fractional transformations not only extend to Clifford algebras, but when reformulated in terms of Clifford algebras provide more insight even into complex numbers— in a word, an elementary presentation of Clifford algebras and their many applications. This subject has become too important to be left in the professionally deformed hands of full-time algebraists.

Similarly, the chapter on p-adic numbers could have been improved had it started out with Ostrowski's theorem, followed by an introduction of the real numbers as a necessary consequence of the theorem. In the present edition, Ostrowski is not even mentioned in the index.

Further, this addition might have aroused some interest in Dedekind cuts—by contrasting the real numbers with what happens in completing the field of rational numbers in the other $\infty - 1$ cases, where a process similar to Dedekind cuts gives the Cantor set instead.

The editors miss another chance later on to introduce a generalization of the concept of number, and that is the ring of idèles. Even the mathematical public is largely unaware of this notion. The ring can be motivated starting with the p-adic numbers, which in turn can (and should) be motivated by the notion of "carrying to the right" while performing additions and multiplications instead of carrying to the left, as in the case of the real numbers. In carrying to the left, it does not matter what base is used, but the fields obtained by carrying to the right are not homeomorphic as topological spaces for different bases p. The ring of idèles sets up a carrying algorithm (to the right) cleverly based on cyclic groups which does not play favorites among the ps. Where else but in a book like this would such a motivation for the ring of idèles be given? Certainly not in a book of algebra, where the author invariably works hard at concealing all fundamental motivations.

Still *à propos* of Cantor, it is not clear why cardinal and ordinal numbers have been dropped in a text that (implicitly) pretends to treat "every" aspect of the concept of number. For example, a description of what is going on in the theory of large cardinals might have made fascinating reading.

We might go on forever pointing out changes that could have improved the contents. More important, however, is the question: will anybody read this book from one end to the other? Will anyone stomach yet another treatment of the real numbers as Dedekind cuts, and of complex numbers as ordered pairs? Does the student benefit from such an uneven presentation? Who cares about the history of π?

We urge the well-meaning but misguided editors to seriously consider a revised edition, to insert a variety of timely and engaging new subjects and mercilessly drop worn-out topics. At a time when mathematics is suffering from a serious loss of status, the circulation of this book among an already fed-up public is another time bomb to be defused — the sooner the better.

On Reading Collected Papers

In our ahistorical age, it is a fool's paradise to believe that the reading (even the casual scanning) of collected papers is more likely to enrich our knowledge than a feverish plunge into the latest periodicals or the rushing from one conference to the next. The masters had a variety of ideas that are missing in later accounts of their work.

The rapid advance of mathematics sifts the men from the boys faster than ever. Like a fragile building in an earthquake, mediocre mathematics falls by the wayside and the sturdy edifices remain. Fashion's paradoxes lead to the victorious comeback of ideas that some of us thought buried forever.

Looking at Emmy Noether's collected works,[5] we are struck by the timeliness of some papers which our teachers had advised us to skip. Her elaboration of Hentzelt's thesis is again the talk of the town, as is her work on differential invariants, now that the *algebra* of differentiation is no less important than the *analysis*. The papers on classical invariant theory make up a good half of her published work, and even her thesis, written at Gordan's suggestion, bears careful scrutiny. It is rumored that the author had a low opinion of her work on invariant theory. We wonder whether she will turn in her grave when some nice young person rewrites it in the jargon of our day. The papers which made her name a household word at St. Algebra's are the least interesting; such are the wages of success.

Van der Waerden[6] has permitted a publisher to string into a book his long series of papers on algebraic geometry. Throughout his life he had been publishing them in careful linear order. Now that we are toilet trained in commutative algebra, we can read them without panic. The more timely ones are the papers that twenty-five years ago would have been pooh-poohed: those on elimination theory, the Schubert calculus and special topics (we are again in an age of the special which manages to survive specialization).

Harish-Chandra[7] is a special case. With him, harmonic analysis came of age in a monument *aere perennius*. Pattern recognition? Image processing? Filtering through noise? Big words in the world of big bucks, problems clamoring for fast solution. The know-nothings will be surprised to learn that the solutions to these "practical" problems are more likely to come from the work of the Harish-Chandras than from the Mickey Mouse animations on which the Federal Government wastes millions. All we need is a team of gifted middlemen, trained in the latest techniques of jungle warfare to carry the message where it is most needed.

Littlewood's collected papers[8] are a shorter set after Hardy's seven volumes. The papers of the two English mathematicians are similar in spirit: they both write with an attention to detail that is characteristic of twentieth century British mathematics. Some of the papers are superseded by later work. Nevertheless, one still reads the originals with interest and often with pleasure. Strangely, one does not feel the need for working out details on a separate sheet of paper. Is it Littlewood's mastery that makes the reading painless?

Dynkin's collection of papers[9] on Markov processes is fresh and live reading in spite of the research which they stimulated, which could have rendered them obsolete.

Magnus' papers[10] display a coherence and unity which make them a delight. We watched with ill-concealed impatience the demise of finite groups in a grand finale. At last combinatorial group theory will be granted an equal place.

Borel[11] is a severe master. His no-nonsense style is more educational than a graduate education, and his papers are closest to our times in spirit as well as content.

Whatever libraries do in their disorderly rush to cut costs, they should stock these and all other sets of collected papers. A university library is a symbol of permanence, and there is no clearer symbol than a voice of the past that keeps us alive and hoping.

Matroids

The subjects of mathematics, like the civilizations of mankind, have finite lifespans, which historians are expected to record.

There are the old subjects, loaded with distinctions and honors. As their problems are solved and their applications enrich engineers and other moneymen, ponderous treatises gather dust in library basements, awaiting the day when a generation as yet unborn will rediscover a forgotten paradise.

Then there are the middle-aged subjects. You can guess them by roaming the halls of Ivy League universities or the Institute for Advanced Study. Their high priests haughtily refuse fabulous offers from provincial universities, while receiving special permission from the President of France to lecture in English at the *Collège de France*. Little do they know that the load of technicalities is critical, soon to crack and submerge their theorems in a layer of oblivion.

Finally, there are the young subjects — combinatorics for instance. Wild-eyed researchers pick from a mountain of intractable problems, childishly babbling the first words of what may turn into a new language. Childhood will end with the first *Séminaire Bourbaki*.

It may be impossible to find a more fitting example of a subject now in its infancy than matroid theory. The telltale signs are an abundance of non-trivial theorems, together with a paucity of coherent theories.

Like many other great ideas of this century, matroid theory was invented by one of the foremost American pioneers, Hassler Whitney. His paper, which remains the best entry to the subject, conspicuously reveals the unique peculiarity of this field, namely, the exceptionally large variety of cryptomorphic definitions of a matroid, each one embarrassingly unrelated to every other and each one harking back to a different mathematical *Weltanschaung*. It is as if one were to condense all trends of present day mathematics onto a single structure, a feat that anyone would *a priori* deem impossible, were it not for the fact that matroids do exist.

The original motivation, both in Whitney and in Tutte's vertiginously profound papers, was graph-theoretic. Matroids are objects that play the role of the dual graph of a graph when the graph is not planar. Almost every fact about graphs that can be formulated without using the term "vertex" has a matroidal analogue. The deepest insight obtained from this yoga is Tutte's homotopy theorem, a unique combinatorial achievement whose far reaching implications are a long way from being fully understood.

But another axiom system for matroids, first adopted by Mac Lane, is Steinitz' axiom. The Mac Lane–Steinitz axiomatization unifies linear independence of vectors and transcendence degree of fields. From the Mac Lane–Steinitz axiom system follows a second yoga: almost every fact about linear or algebraic independence is a theorem about matroids. For example, Higgs was able to develop a matroidal analogue of a linear transformation.

The conjecture inevitably followed that matroids could be represented by points in projective space. But this conjecture was quickly quashed by Mac Lane. The conditions for representability of a matroid as a set of points in some projective space can be stated in terms of the absence of "obstructions" or forbidden configurations, as in the archetypal theorem of Kuratowski characterizing planar graphs. Such theorems are difficult to come by; witness Seymour's theorem characterizing all matroids that can be represented as sets of points in a projective space over a field with three elements.

Next, we find the lattice theorists — Birkhoff, Crapo — whose yoga is to visualize a matroid by its lattice of flats. Deep enumerative properties, often associated with homological properties, hold for the lattice of flats of a matroid, now known as a geometric lattice. The Whitney numbers of a geometric lattice (a term introduced by Harper and Rota) exhibit some of the properties of binomial coefficients, as Dowling and Wilson proved, though not all, as a counterexample of Dilworth and Greene showed.

As if three yogas were not enough, there followed the Tutte-Grothendieck decomposition theory (developed most energetically by Tommy Brylawski), which displayed an astonishing analogy with *K*-

theory. Zaslawsky's solution to Steiner's problem, which gives an explicit formula for the number of regions into which space is subdivided by a set of hyperplanes, has blossomed into the theory of arrangements of hyperplanes, a current yoga.

What else? Kahn and Kung came along with a new yoga: a varietal theory of matroids that brings universal algebra into an already crowded game. Seymour's decomposition theory will probably lead to yet another yoga.

All these yogas arouse a suspicion. Anyone who has worked with matroids has come away with the conviction that matroids are one of the richest and most useful mathematical ideas of our day. Yet we long, as we always do, for *one* idea that will allow us to see through the variety of disparate points of view. Whether this idea will ever come along will depend largely on who reads the essays collected in this fine volume.[12]

Short Book Reviews

Z. W. Pylyshyin, *Computation and Cognition*, MIT Press: Cambridge, MA, 1984.

I once asked a biologist friend of mine who works on the genetics of DNA to describe his work. "It is like trying to infer the workings of the Federal Government from an inspection of the buildings in Washington," she said. It occurred to me that the same problem arose once in the phenomenology of perception, as the problem of *Fundierung*; now quashed by the avalanche of reductionism that envelops the life sciences. *Quos Deus vult perdere, dementat prius.*

S. Albeverio, G. Casati, D. Merlini, *Stochastic Processes in Classical and Quantum Systems*, Springer, New York, 1986.

Ah, the Ticino school of combinatorics! Ah, the beautiful new stochastic processes it is creating from the fusion of probability and quantum mechanics! Ah, how we wish the day will come when these new stochastic processes will find an interpretation in the real world!

B. E. Sagan, *The Symmetric Group*, Wadsworth: Monterey, CA, 1991.

A responsible, readable, rational, reasonable, romantic, rounded, respectable, remarkable repertoire of results on a range that has rarely been so rightly reorganized.

W. Linde, *Probability in Banach Spaces — Stable and Infinitely Divisible Distributions*, Wiley, New York, 1986.

No kidding: the theory of Banach spaces, several times declared defunct, is now finding that probability is its latest customers. For a while it looked like Banach spaces were turning into another chapter of *Desperationsmathematik*, with rearrangements of series and other odd curios from the glass menagerie of the nineteenth century. But let us not forget: good mathematics will always find good applications. Just wait long enough. This statement of faith is not just an application of the ergodic theorem.

E. Husserl, *Vorlesungen über Bedeutungslehre*, Nijhoff: The Hague, 1987.

Compared to Husserl's, the paradise that Cantor bequeathed to us is a limbo. It says in the Bahgavad Gita: "When you me fly, I be the wings."

J. H. Conway, N. J. A. Sloane, *Sphere Packings, Lattices and Groups*, Springer, New York, 1988.

This is the best survey of the best work in one of the best fields of combinatorics, written by the best people. It will make the best reading by the best students interested in the best mathematics that is now going on.

Allan Bloom, *The Closing of the American Mind*, Simon and Schuster, New York, 1987.

One of the irrepressible mistakes of our time is the zero/one mistake, as we like to call it. It occurs when we take one single failing in an author's work and proceed to infer that all his work must be no good; or conversely, when we decide in advance that an author's work must be good, because the author is a good guy— by our standards, of course. Bloom's *The Closing of the American Mind*, a bestseller a few years ago already all but forgotten, is a case in point. Admitting that SOME of his assertions are true and SOME others are nonsensical takes courage. Few will undertake the effort to separate the wheat from the chaff.

T. E. Cecil, *Lie Sphere Geometry*, Springer, New York, 1992.

Everyone is aware of the fact that fashions in mathematics are coming in and going out with the speed of a revolving door in rush hour. But perhaps not many have yet to notice a healthy consequence of accelerated obsolescence: the forgotten mathematics of the nineteenth century is given a new lease on life, and is coming back into fashion with a vengeance. The dead are rising from their graves instead of turning in them, for a change.

S. K. Campbell, *Flaws and Fallacies in Statistical Thinking*, Prentice-Hall, Englewood Cliffs, NJ, 1974.

It is well known that statistics is that branch of applied mathematics that is expressly designed to reinforce conclusions previously decided upon. The author seems to believe that this fundamental role of statistics in our culture is an oddity rather than a service.

C. S. Hardwick, *Language Learning in Wittgenstein's Later Philosophy*, Mouton, The Hague, 1971.

Rumor has it that when Ludwig Wittgenstein died, a worn and marked-up copy of Heidegger's *Being and Time* was found in his quarters at Cambridge University. The writings of the later Wittgenstein arouse the suspicion that the author's purpose was to provide examples for Heidegger's example-free phenomenology.

C. S. Hornig, *Volterra-Stieltjes Integral Equations*, North-Holland, Amsterdam, 1975.

Certain applied mathematicians like to allude to "applications" with the same awed breathlessness as a bureaucrat would refer to "the office upstairs" in a Kafka novel. Whenever there is too much talk of applications, one can rest assured that the theory has very few of them.

W. V. Quine, *Ontological Relativity*, Columbia University Press, New York, 1966.

When a philosopher writes well, one can forgive him anything, even being an analytic philosopher.

Serge Lang, *Cyclotomic Fields*, Springer, New York, 1978; *Cyclotomic Fields II*, Springer, New York, 1980; *Introduction to Modular Forms*, Springer, New York, 1976.

Some day, when we are all gone, one name will be remembered and revered, that of Serge Lang. He has realized that mathematics cannot survive without the synthetic view that pulls disparate threads together and gives the big picture. With his expositions of fields that nobody would otherwise write up he has done more for the advancement of mathematics than anyone now alive. In so doing, he has met the fatuous criticism of those who are wedded to the "one-shot" view of mathematics. According to this view, mathematics would consist of a succession of targets, called problems, which mathematicians would be engaged in shooting down by well-aimed shots. But where do problems come from, and what are they for? If the problems of mathematics were not instrumental in revealing a broader truth, then they would be indistinguishable from chess problems or crossword puzzles. Mathematical problems are worked on because they are pieces of a larger puzzle. The solution of individual problems is valuable only insofar as it serves to build a theory. But it takes cultivation, a rare commodity in our day, to see all this.

O. Teichmüller, *Gesammelte Abhandlungen — Collected Papers*, Springer, Berlin, 1982.

Teichmüller was one of the mathematical geniuses of the twentieth century. He was an unpleasant person in some ways, and few would agree nowadays with his political views (it is no longer fashionable to be an Obersturmführer in the SS). His mathematical papers still make interesting reading. It is discouraging to note time and again how talent for mathematics is independent of sound judgment. We would like to be able to say that a great mathematician is also a great man. But reality is seldom what it ought to be, and the greater the mathematicians, the greater the failings (to paraphrase Heidegger). Fortunately, we have biographers whose job is to present the life that ought to have been and to allow us to forget reality without regrets.

J. F. Hofmann, *Leibniz in Paris*, Cambridge University Press, Cambridge, 1974.

The author was one of the foremost experts ever on Leibniz, and this is his life's work. Unlike most historians of today, he writes engagingly and accessibly. This book should go a long way to eliminate the perniciously inaccurate romantic image of the superior-to-all, universal, saintly "genius," an image which is still inculcated, with criminal disregard for the truth and catastrophic results, into schoolchildren all over the world.

J. Neukirch, *Class Field Theory*, Springer, New York, 1986.

At last, an exposition of class field theory that does not presume of the reader full knowledge of several not yet written (and never to be written) books. The chilling elegance of this presentation will give you goose pimples.

E. Husserl, *Phantasie, Bewusstsein, Erinnerung*, Nijhoff, The Hague, 1980.

The late Edmund Husserl, a student of Weierstrass turned philosopher whom Gödel thought to be the greatest philosopher since Leibniz, is still publishing, fifty years after his death, at the rate of one volume per year. Too bad he can no longer be considered for tenure.

J. L. Mackie, *Truth, Probability and Paradox*, Oxford University Press, Oxford, 1973.

One more book that treats the problems of two generations ago. The chapter on probability seems written by someone living on the moon: no account of the current revival of von Mises' *Kollektif*; only a perfunctory discussion — inferior to Sir Harold Jeffrey's — of the Bayesian point of view, and total silence on the role of probability in quantum mechanics. As meager compensation, we are given the usual trite list of paradoxes. What can we expect of a writer who prefaces his book with the bigoted statement that "all philosophy, to be any good, must be analytic?"

J. L. Mackie, *The Cement of the Universe: A Study of Causation*, Oxford University Press, New York, 1974.

The ineffable Mr. Mackie is at it again. He gives us his pretty world where facts are expressed by statements of the form "A is B," where events have definite causes and effects, where everything is made of atoms and molecules. The world of our infancy, in short. A world we would like escape into and take shelter in, away from the messy uncertainties of today. But alas, you can't go home again.

J. Passmore, *Recent Philosophers*: a supplement to *A Hundred Years of Philosophy*, Open Court, London, 1985.

When pygmies cast such long shadows, it must be very late in the day.

Acknowledgments

Chapter I. Reprinted with permission of the American Mathematical Society.

Chapter II. Reprinted with permission of Springer Verlag.

Chapter IV. Reprinted with Permission of Letters in Mathematical Physics.

Chapter VII. Reprinted with permission of the Journal of Metaphysics and Synthèse.

Chapter IX. Reprinted with permission of The Mathematical Scientist.

Chapter XII. Reprinted with permission of Philosophy of Science Society.

Chapter XIV. Reprinted with permission of Marquette University Press

Chapter XV. Reprinted with permission of The Hegeler Institute.

Chapter XVI. Reprinted with permission of Kluwer Academic.

Chapter XIX. Reprinted with permission of The Mathematical Association of America.

Chapter XXI. Reprinted with permission of The Mathematical Association of America and Academic Press.

End Notes

by Fabrizio Palombi

I. Fine Hall in its Golden Age

[1] Alonzo Church, *Introduction to Mathematical Logic, I*, Princeton University Press, Princeton, 1956

[2] Bertrand Russell, Alfred North Whitehead, *Principia Mathematica*, Cambridge University Press, Cambridge, 1925–27

[3] Alonzo Church, Conditioned Disjunction as a Primitive Connective for the Propositional Calculus, *Portugalie Mathematica*, no. 7, 1948, 87–90

[4] William Feller, *An Introduction to Probability Theory and its Applications*, John Wiley & Sons Inc., New York, 1950

[5] George Pólya, Gabor Szegö, *Aufgaben und Lehrsätze aus der Analysis*, 2 vols., Springer Verlag, Berlin,1925

II. Light Shadows

[1] Ralph Waldo Emerson, *Representative Men*, 1850

[2] Erwin Chargaff, *Heraclitean Fire*, Rockfeller University Press, New York, 1978

[3] Nelson Dunford, Integration in General Analysis, *Transaction of the American Mathematical Society*, vol. XXXVII, 1935, 441–453

[4] Nelson Dunford, Jacob T. Schwartz, *Linear Operators, Vol. I*, Interscience, New York, 1958; *Vol. II* , Wiley, New York, 1963; *Vol. III*, Wiley, New York, 1971

[5] Alexandre Grothendieck, Résumé de la théorie métrique des produits tensoriels topologiques, *Boletin de Sociedade de Matematica de São-Paulo*, no. 8, 1956, 1–19

III. The Story of a Ménage à Trois

[1] Alfred Young, *Collected Papers*, University of Toronto Press, Toronto, 1970

[2] Hermann Weyl, *Gruppentheorie und Quantenmechanik*, Hirzel, Leipzig, 1928

[3] Van der Warden, *Moderne Algebra, I* Teil, 1937; II Teil, 1940, Springer, Berlin

[4] Hermann Grassman, *Hermann Grassman's gesammelte mathematische und physikalische Werke*, 6 Vol., Druck und Verlag von B.G., Teubner, 1894–1911

[5] Eduard Study, *Einleitung in die Theorie der Invarianten linear Trasformationen auf Grund der Vektorenrechnung*, I Teil, Vieweg, Braunschweig, 1923

[6] Giuseppe Peano, *Calcolo geometrico secondo l'Ausdehnungslehre di Grassmann*, Bocca, Torino, 1888

[7] Luther Pfahler Eisenhart, *Introduction to Differential Geometry with use of the Tensor Calculus*, Princeton University Press, Princeton, 1947

[8] Imre Lakatos, *Proofs and Refutations*, Cambridge University Press, Cambridge, 1976

[9] Nicholas Bourbaki, *Eléments de mathématique. Première partie: algèbre. Chapitre 3*, Hermann, Paris, 1948

[10] F. N. David, D. E. Barton, *Combinatorial Chance*, Griffin, London, 1962

VI. The Lost Café

[1] Stefan Banach, *Théorie des opérations linéaires*, Subwencji funduszu Kultury naradoweg, Warszawa, 1932

[2] John Oxtoby, Stanislaw Ulam, Measure-preserving Homeomorphisms and Metrical Transitivity, *Annals of Mathematics*, vol. XLII, 874–920

[3] Charles J. Everett, Stanislaw Ulam, Projective Algebra I, *American Journal of Mathematics*, vol. LXVIII, 1946, 77–88

VII. The Pernicious Influence of Mathematics Upon Philosophy

[1] J. T. Schwartz, "The Pernicious Influence of Mathematics Upon Science" in Mark Kac, Gian-Carlo Rota, J. T. Schwartz, *Discrete Thoughts. Essays on Mathematics, Science, and Philosophy*, Birkhäuser Boston, 1992, Chapter 3

[2] Ludwig Wittgenstein, *Tractatus logico-philosophicus*, Kegan, Trench, Trubner & Co. Ltd., London, 1922; Proposition 7.00

[3] Ludwig Wittgenstein, *Philosophische Untersuchungen*, Basil Blackwell, Oxford, 1953

VIII. Philosophy and Computer Science

[1] Eugenio Montale, *Ossi di seppia*, in *Opere*, Mondadori, Milano, 1984

IX. The Phenomenology of Mathematical Truth

[1] Norman Levinson, On the Elementary Proof of the Prime Number Theorem, *Proceedings of the Edinburgh Mathematical Society*, (2) 15, 1966–67, 141–146

X. The Phenomenology of Mathematical Beauty

[1] G. F. Hardy, *A Mathematician's Apology*, Cambridge University Press, Cambridge, 1967

[2] David Hilbert, *Die Grundlagen der Geometrie*, 7th ed., B. G. Teubner, Leipzig, 1930

[3] Emil Artin, *Galois Theory*, Notre Dame Mathematical Expositions, University of Notre Dame, 1941

[4] David Hilbert, Die Theorie der Algebraischen Zahlkörper, *Jahresbericht der Deutschen Mathematikvereinigung*, Vol. IV, 175–546

[5] H. Weber, *Lehrbuch der Algebra*, 3 vols, Braunschweig, Vieweg, 1895–96

XI. The Phenomenology of Mathematical Proof

[1] Ronald L. Graham, Bruce L. Rothschild, Joel H. Spencer, *Ramsey Theory*, Wiley, New York, 1990

[2] Bertram Kostant, The principle of Triality and a distinguished unitary representation of $SO(4,4)$, *Geometric Methods in Theoretical Physics*, K. Bleuler and M. Werner (eds.), Kluwer Academic Publishers, 1988, 65–108;
The Coxeter Element and the Structure of the Exceptional Lie groups, Colloquium Lectures of the AMS, Notes available from the AMS.

[3] Andrew Wiles, Modular Elliptic Curves and Fermat's Last Theorem, *Annals of Mathematics*, Vol. CXLII, 1995, 443–551

[4] H. F. Baker, *Principles of Geometry. Vol. I, Foundations,* Cambridge University Press, Cambridge, 1922

[5] G. D. Birkhoff, Proof of the Ergodic Theorem, *Proceedings of the National Academy of Sciences*, Vol. XVII, 1931, 656-660

[6] Adriano M. Garsia, A Simple Proof of E. Hopf's Maximal Ergodic Theorem, *Journal of Mathematics and Mechanics*, Vol. XIV, 1965, pp. 381–382

[7] Hans Lewy, On the local character of the solutions of an atypical differential equation in three variables and a related problem for regular functions of two complex variables, *Annals of Mathematics*, Vol. LXIV, 1956, pp. 514-22

[8] von Staudt, *Geometrie der Lage*, 1847

[9] Garrett Birkhoff, *Lattice Theory*, American Mathematical Society, 1948

[10] Emil Artin, *Coordinates in Affine Geometry*, Reports of Mathematical Colloquium, Notre Dame, 1940

XII. Syntax, Semantics and the Problem of the Identity of Mathematical Items

[1] Article written in collaboration with David Sharp and Robert Sokolowski

XIII. The Barber of Seville

[1] Gian-Carlo Rota, David Sharp, Robert Sokolowski. This paper is a

response to a letter of Professor Spalt in which the content of the previous chapter was criticized

XV. Fundierung as a Logical Concept

[1] Edmund Husserl, *Logische Untersuchungen*, Halle, M. Niemeyer, 1900–1

[2] Ludwig Wittgenstein, *Philosophische Untersuchungen*, Oxford, Basil Blackwell, 1953

[3] Gilbert Ryle, *Dilemmas, The Tarner Lectures*, Cambridge University Press, Cambridge,1954

[4] John.L. Austin, *Sense and Sensibilia*, Oxford University Press, London, 1962

XVI. The Primacy of Identity

[1] Robert Sokolowski, *Pictures, Quotations and Distinctions. Fourteeen Essays in Phenomenology*, University of Notre Dame Press, 1992
Robert Sokolowski, *Edmund Husserl and the Phenomenological Tradition*, Studies in Philosophy and in the History of Philosophy, The Catholic University of America Press, Washington, 1988
Robert Sokolowski, *Moral Action. A Phenomenological Study*, Indiana University Press, Bloomington, 1978
Robert Sokolowski, *Husserlian Meditations*, Northwestern University Press, Evanston, 1974
Robert Sokolowski, *The Formation of Husserl's Concept of Constitution, Phenomenologica 18*, Martinus Nijhoff, The Hague, 1964

[2] The New York Times Book Review, September 11, 1994, page 31.

XVII. Three Sense of "A is B" in Heidegger

[1] Martin Heidegger, *Gesamtausgabe*, Band 1, *Frühe Schriften*, Klostermann, Frankfurt am Main, 1978, p. 42 and p. 174

[2] Martin Heidegger, *Gesamtausgabe*, Band 2, *Sein und Zeit*, Klostermann, Frankfurt am Main, 1977

[3] Martin Heidegger, *Gesamtausgabe*, Band 65, *Beiträge zur Philosophie*, Klostermann, Frankfurt am Main, 1989

Book Reviews

[1] Paul R. Halmos, *I Want to Be a Mathematician: an Automathography*, Springer, New York, 1985
[2] Schaum's Outlines Series in Mathematics and Statistics, McGraw-Hill Publishing Company, New York
[3] T.Y. Lam, *First Course in Noncommutative Rings*, Springer, New York, 1991
[4] H. D. Ebbinghaus, H. Hermes, F. Hirzebruch, M. Koecher, K. Mainzer, J. Neukirch, A. Prestel, and R. Remmert (Eds.), *Numbers*, Springer, New York, 1991
[5] E. Noether, *Gessammelte Abhandlungen*, Springer, Heidelberg, 1983
[6] B. L. van der Waerden, *Zur algebraischen Geometrie (Selected Papers)*, Springer, Heidelberg, 1983
[7] Harish-Chandra, *Collected Papers*, 4 vols., Springer, Heidelberg, 1982
[8] J. E. Littlewood, *Collected Papers*, 2 vols., Oxford University Press, Oxford, 1982
[9] E. B. Dynkin, *Markov Processes and Related Problems of Analysis (Selected Papers)*, Cambridge University Press, Cambridge, 1982
[10] Wilhelm Magnus, *Collected Papers*, Springer, Heidelberg, 1984
[11] A. Borel, *Collected Papers*, 3 vols., Springer, Heidelberg, 1983
[12] P. S. Kung, *A Source Book in Matroid Theory*, Birkhäuser, Boston, 1986.

Epilogue

[1] Ester Rota Gasperoni, *L'orage sur le lac*, Médium, Paris, 1995
[2] Ester Rota Gasperoni, *L'arbre des capulíes*, Médium, Paris, 1996
[3] Alain, *Propos*, vol. 1, Editions de la Pléiade, Gallimard, Paris, 1956, p. 467
[4] José Ortega y Gasset, *Obras Completas*, vol. V, Revista de Occidente, Madrid, 1955, pp. 379-410
[5] José Ortega y Gasset, *Obras Completas*, vol. IV, Revista de Occidente, Madrid, 1951, p. 32

Epilogue

Fabrizio Palombi

As the editor of this book, I feel it is my duty to remark on a number of glaring issues in the preceding text.

First of all, a few words about the author. I have often asked Rota how he can manage to live in two disparate worlds, the world of combinatorics and the world of phenomenology. His answers are invariably evasive. One that I find exasperating is, "There is no answer to the 'how.' You do not need an answer to your question, you need instead to acknowledge the fact that I am that way."

If I were asked which of these two strange worlds Rota is fonder of, I would opt for the world of phenomenology. Rota is a phenomenologist manqué. I cannot understand his mathematics, but I suspect mathematics is for him a substitute for phenomenology. Rota is a little too fond of repeating that philosophy in our time has fallen to its lowest depth since the seventh century A.D. Perhaps he is right. Yet Rota does not seem uncomfortable with his philosophical isolation. He is too uncompromising towards friends and colleagues who make an effort to understand phenomenology, but do not quite "get it." Very few people are able to understand phenomenology. The basic message of phenomenology is simple, but far too revolutionary, and most people reject it out of fear.

I have a number of stock answers I use when asked the question, "What is phenomenology?" the most shocking of which I probably heard from Rota. It goes without saying that I had to tone down Rota's language, as in the sentence: "Phenomenology is the formalization of context dependence." For some time I believed this answer comes close to the truth, but one day Rota destroyed it for me. After singing the praises of "formalization" for umpteen years, one day without any warning he turned around and started claiming that no one knows what the word "formalization" means. It was back to square one for me.

As a backup, I picked up another of his "definitions" of phenomenology: "Phenomenology is the most extreme form of realism." Again, I made use of this second definition for a while to get rid of embarrassing questioners, until one day I realized — this time without Rota's help — that I did not have an inkling of what the word "realism" means. How do you convince a hardened materialist who believes that the only reality there is is the reality of atoms and molecules that he is not being realistic? How do you find his Achilles' heel? What is the answer to *this* question, Mr. Rota?

Rota told me one day that people are inexorably subdivided into two kinds: materialists and transcendentalists. If you are born a materialist, there is nothing phenomenology can do for you. No amount of phenomenological argument will convince you of the absurdity of materialism. So far, so good. Next, he turns against the transcendentalists. If you are born a transcendentalist, you are likely to find phenomenology to be the statement of trivialities. Does this mean that you do not need phenomenology? "Hardly," says Rota. "Phenomenology must be used to stop the transcendentalist from becoming overly transcendental." This is all fine and dandy, but it contradicts Rota's claim that phenomenology reached its peak with Heidegger. As a matter of fact, he himself has complained to me about Heidegger's half-serious, repetitious, and constantly self contradicting style. He tells me one must see through these quirks of Heidegger's. Tell me about it.

Rota maintains that Husserl is the best phenomenologist, and I agree. Husserl's brass tack descriptions come as close to realism as anything I have read. Husserl's distinction of several kinds of *Fundierung* is typical. Rota told me that he has deliberately ignored these distinctions, I don't know why. Rota quotes Lebesgue as stating that every mathematician must be something of a naturalist. Husserl said pretty much the same thing about philosophers, way before Lebesgue. Every philosopher must from time to time engage in detailed description, whether phenomenological or not. Descriptions of special cases are the nitty-gritty of philosophy. And teaching the born transcendentalist the art of phenomenological description will stop him from going wild, it will keep him chained to reality.

Rota must indeed have felt chained to reality during World War II. His anti-Fascist father was condemned to death by the late Duce, and saved his skin by crossing over the Alps to Switzerland. His sister has written up and published in a two-volume set[1,2] the hard-to-believe story of his family's wanderings in the Alps during World War II, while their father was being sought by the police. Rota had never told me a word of it, until one day he asked me out of the blue to drive him from my native Novara to some forlorn hamlet on the shore of Lake Orta. He then told me, while waving at the lake, how in the winter of 1944 he used to cross the lake with other children to go to school from the Partisan-held side of the lake to the Fascist-held side, with a white flag on top of the boat. I had trouble believing him, until he roughed me up with some words in dialect. He can still speak the ancient dialect of his native Vigevano (which is almost identical to the dialect of my Novara). Nowadays almost no one speaks it, and every time he shows up in Vigevano he speaks it as if he had never left. But as a matter of fact he has not lived in Italy since 1946, except for brief periods lasting no more than one month. There must be a genetic explanation for the permanence of these old dialects which can only be learned in the cradle.

Rota's experiences in World War II must have predisposed him for philosophy. Listen to this definition he gave of phenomenology, the one I find most controversial: "Phenomenology is the study of three words: as, already and beyond." Enter Heidegger again. It brings out one of Rota's three philosophical "obsessions": the obsession with what Heidegger calls the *Augenblick* (actually, Heidegger uses several terms for the same concept. This does not help the reader.) Rota claims that Heidegger's theory of the *Augenblick* can be found in Husserl's sixth logical investigation; he insists that all of Heidegger's philosophy is a rewrite of Husserl's sixth logical investigation in a more attractive language. This preference of Rota for Heidegger's style over Husserl's runs counter to his background as a mathematician. Husserl writes in the style of Weierstrass, his thesis advisor, whereas while reading Heidegger you often hear Kierkegaard whining in the background. Rota imitates Heidegger's style in "Three Senses of 'A is B' in Heidegger," with what success the reader can judge. I doubt that any reader who is not a dyed-in-the-wool Heideggerian can understand his chapter. His summary of Husserl's main thesis in "The Primacy of Identity" is also hard to read. The fundamental assertion that identity precedes existence is hard to swallow after such condensed presentation.

One of the points on which I most heartily agree with Rota is that Heidegger blew Husserl's sixth logical investigation out of all proportion in his "Being and Time." Someone other than Heidegger ought to rewrite the sixth logical investigation so that the average cultivated reader of today can understand. Rota complains that philosophers are not in the habit of improving upon their results by rewriting them, unlike mathematicians.

Now you will ask: what are Rota's other two obsessions? I am glad to comply with your request. One of them is the obsession with split personality. His descriptions of mathematicians are meant to bring out contradictory traits that most people prefer to forget. I myself like a good mathematician to be a good guy. Here comes Rota and tells us that Lefschetz threw into the wastebasket manuscripts received for publication in *Annals of Mathematics* and then claimed never to have received them, and that Emil Artin twisted his students' personalities until they became carbon copies of his own. It is sad to admit that great men have flawed personalities, as Rota keeps reiterating. Why can't we go on doing what has been done since Plutarch, embellishing history and writing hagiographies?

If Aristotle only knew how his theory of the unity of personality is being torn to shreds !

The strongest of Rota's chapters is "The Lost Café." What made him single out Stan Ulam's foibles for extensive description? Doesn't he know that one does not say this kind of thing about great men? He has had to bear the consequences of his actions. Mrs. Ulam made sure that he learned his lesson. She personally crossed out his name from the list of participants at every meeting that has taken place in the memory of Ulam.

After iterating his message of split personalities and contradictory traits, Rota makes a quick turnaround and professes to dislike psychoanalysis. He rants and raves against Sigmund Freud, quoting Nabokov, who called him "that Viennese con man and crackpot." Doesn't Rota realize that without Freud's spadework he could not even dream of twisted personality?

But let us go on to the third obsession in Rota's philosophy: the obsession with the word "existence." He believes that most discussions of this word are motivated by emotional cravings ("emotional cravings" is one of his favorite expressions), and are devoid of philosophical content, indeed of any content whatsoever. His most vicious attack deals with the "existence" of mathematical items. It is found in the chapter "The Barber of Seville, or the Useless Precaution," a rude letter sent to a well-meaning and dignified German philosopher of science, signed by Sokolowski and Sharp as well as Rota. In this letter Rota begins to make fun of those honest and hard working folks who hold on to some firm notion of reality. Wouldn't it be better all around if the items of mathematics "existed" in some sense or other? Why not give them this little satisfaction? Why deprive people of their myths? But deprive them he does with vicious glee in the chapters that follow the letter to Spalt. Doesn't he realize that emotional considerations are to be respected in philosophical argument? We *need* to believe that the world "exists"; the word "existence" may be meaningless outside physics and mathematics, a point that Rota brutally hammers in, but the word "existence" is *emotionally* indispensable.

By far the most controversial of Rota's chapers is "The Pernicious Influence of Mathematics Upon Philosophy." After the original article was published in *Synthèse*, my analytic philosopher friends came to me one after another and told me: "Look what your friend is writing about us!" I had trouble calming them down by telling them that Rota did not really mean to attack them, that it was all a joke.

Another chapter that needs some comment is "The Barrier of Meaning." Rota has assured me that the conversation with Stan Ulam described in this chapter really did take place, but I have some doubts.

Rota's critical mind is in full swing in his three chapters on the philosophy of mathematics. In "The Phenomenology of Mathematical Truth" he gives two conflicting theories of mathematical truth, without choosing between the two. This is a heavy hint to the reader that both theories have had their day. A similar trick is played upon the reader in the next chapter, "The Phenomenology of Mathematical Beauty." Rota claims that beauty is "enlightenment," but he deliberately stops short of explaining what is meant by "enlightenment." In the next chapter, "The Phenomenology of Mathematical Truth," we find a similar rhetorical device. Again he dangles before the reader one after the other notion of mathematical proof, only to take them back as soon as the reader begins to take them seriously.

Rota's texts are rich in hidden references and ironies which only a well read reader will catch. I feel I must give some samples. The section "Problem Solvers and Theorizers" in the chapter "The Story of a Ménage à Trois" is an imitation of an essay by Alain[3] which is titled "Janséniste et Jésuite." Another instance: the introduction he wrote is inspired by an essay by Ortega y Gasset (*Ideas y creencias*, in the fifth volume of his collected papers.[4]) The title "The Barber of Seville, or the Useless Precaution" is taken from Ortega's essay on Kant[5].

I also noticed that Rota's quotations from Hegel have been filtered through Croce, one of our Italian philosophers whom Rota mentions only once in the whole book, even though he uses Croce's ideas repeatedly. Rota's philosophical development has also been strongly influenced by John Dewey, a philosopher whom he extensively read in his youth.

In closing, I cannot hide my suspicion about the sentence about pygmies, which is the closing sentence of the book. I was at first convinced that this sentence had been lifted out of Chargaff (it is prime-time Chargaff, and Rota sings the praises of Chargaff's autobiography

Heraclitean Fire), and I was ready to add a footnote to that effect, but to my chagrin I could not locate the sentence in any of Chargaff's writings.

I wish Rota had included some of the hate mail he has received over the years for his book reviews. *That* would make interesting reading, for a change!

Novara, August 31, 1996

Index